Learning Functional Data Structures and Algorithms

Learn functional data structures and algorithms for your applications and bring their benefits to your work now

Atul S. Khot
Raju Kumar Mishra

BIRMINGHAM - MUMBAI

Learning Functional Data Structures and Algorithms

First published: February 2017

Production reference: 1170217

Published by Packt Publishing Ltd.

Livery Place

35 Livery Street

Birmingham B3 2PB, UK.

ISBN 978-1-78588-873-1

www.packtpub.com

Credits

Authors
Atul S. Khot
Raju Kumar Mishra

Reviewer
Muhammad Ali Ejaz

Commissioning Editor
Kunal Parikh

Acquisition Editor
Denim Pinto

Content Development Editor
Nikhil Borkar

Technical Editor
Hussain Kanchwala

Copy Editor
Gladson Monteiro

Project Coordinator
Sheejal Shah

Proofreader
Safis Editing

Indexer
Mariammal Chettiyar

Graphics
Abhinash Sahu

Production Coordinator
Shantanu Zagade

About the Authors

Atul S. Khot grew up in Marathwada, a region of the state of Maharashtra, India. A self-taught programmer, he started writing software in C and C++. A Linux aficionado and a command-line guy at heart, Atul has always been a polyglot programmer. Having extensively programmed in Java and dabbled in multiple languages, these days, he is increasingly getting hooked on Scala, Clojure, and Erlang. Atul is a frequent speaker at software conferences, and a past Dr. Dobb's product award judge. In his spare time, he loves to read classic British detective fiction. He is a foodie at heart and a pretty good cook. Atul someday dreams of working as a master chef, serving people with lip-smacking dishes.

He was the author of *Scala Functional Programming Patterns* published by Packt Publishing in December 2015. The book looks at traditional object-oriented design patterns and shows how we could use Scala's functional features instead.

I would like to thank my mother, late Sushila S. Khot, for teaching me the value of sharing. Aai, I remember all you did for the needy girl students! Your support for the blind school - you brought hope to so many lives! You are no more, however your kindness and selfless spirit lives on! I know you are watching dear mother, and I will carry on the flame till I live! I also would like to thank my father, late Shriniwas V. Khot. Anna, I have a photo of the 'Tamra pat'--an engraved copper plaque--commemorating your great contribution to the country's freedom struggle. You never compromised on core values --always stood for the needy and poor. You live on in my memories--a hero forever! I would also want to thank Martin Odersky for giving us the Scala programming language. I am deeply thankful to Rich Hickey and the Clojure community for their work on persistent data structures. Chris Okasaki's "Purely Functional Data Structures" is a perennial source of inspiration and insight. I wish to thank Chris for writing the book. This book is influenced by many ideas Chris presented in his book. I also wish to thank the functional programming community for all the technical writings which is a source of continual learning and inspiration. I would love to express my heartfelt thanks to Nikhil Borkar for all the support through out the book writing. I also would take this opportunity to thank Hussain Kanchwala for the detailed editing efforts to make the book perfect. You guys are awesome! Thanks to y'all!

Raju Kumar Mishra is a consultant and corporate trainer for big data and programming. After completing his B.Tech from Indian Institute of Technology (ISM) Dhanbad, he worked for Tata Steel. His deep passion for mathematics, data science, and programming took him to Indian Institute of Science (IISc). After graduating from IISc in computational science, he worked for Oracle as a performance engineer and software developer. He is an Oracle-certified associate for Java 7. He is a Hortonworks-certified Apache Hadoop Java developer, and holds a Developer Certification for Apache Spark (O'Reilly School of Technology and Databriks), and Revolution R Enterprise-certified Specialist Certifications. Apart from this, he has also cleared Financial Risk Manager (FRM I) exam. His interest in mathematics helped him in clearing the CT3 (Actuarial Science) exam.

My heartiest thanks to the Almighty. I would like to thank my mom, Smt. Savitri Mishra, my sisters, Mitan and Priya, and my maternal uncle, Shyam Bihari Pandey, for their support and encouragement. I am grateful to Anurag Pal Sehgal, Saurabh Gupta, and all my friends. Last but not least, thanks to Nikhil Borkar, Content Development Editor at Packt Publishing for his support in writing this book.

About the Reviewer

Muhammad Ali Ejaz is currently pursuing his graduate degree at Stony Brook University. His experience, leading up to this academic achievement, ranges from working as a developer to cofounding a start-up, from serving in outreach organizations to giving talks at various prestigious conferences. While working as a developer at ThoughtWorks, Ali cofounded a career empowerment based start-up by providing photographers a platform to showcase their art and be discovered by potential employers. His passion for computer science is reflected in his contributions to open source projects, such as GoCD, and his role in serving as the cofounder and Community Outreach Director of a non-profit organization, "Women Who Code - Bangalore Chapter". Along with this, he has also been given the opportunity to speak at different conferences on Continuous Integration and Delivery practices.

When he is not coding, he enjoys traveling, reading, and tasting new cuisine. You can follow him on Twitter at @mdaliejaz.

I want to thank my Mom and Dad, who have always been my inspiration. I'd also like to thank Ahmad and Sana, my siblings, who have been a constant source of cheerful support. A lot of what I am today is because of them.

www.PacktPub.com

For support files and downloads related to your book, please visit www.PacktPub.com.

Did you know that Packt offers eBook versions of every book published, with PDF and ePub files available? You can upgrade to the eBook version at www.PacktPub.com and as a print book customer, you are entitled to a discount on the eBook copy. Get in touch with us at service@packtpub.com for more details.

At www.PacktPub.com, you can also read a collection of free technical articles, sign up for a range of free newsletters and receive exclusive discounts and offers on Packt books and eBooks.

https://www.packtpub.com/mapt

Get the most in-demand software skills with Mapt. Mapt gives you full access to all Packt books and video courses, as well as industry-leading tools to help you plan your personal development and advance your career.

Why subscribe?

- Fully searchable across every book published by Packt
- Copy and paste, print, and bookmark content
- On demand and accessible via a web browser

Customer Feedback

Thank you for purchasing this Packt book. We take our commitment to improving our content and products to meet your needs seriously—that's why your feedback is so valuable. Whatever your feelings about your purchase, please consider leaving a review on this book's Amazon page. Not only will this help us, more importantly it will also help others in the community to make an informed decision about the resources that they invest in to learn. You can also review for us on a regular basis by joining our reviewers' club. **If you're interested in joining, or would like to learn more about the benefits we offer, please contact us**: customerreviews@packtpub.com.

Table of Contents

Preface

This book is about functional algorithms and data structures. Algorithms and data structures are fundamentals of computer programming.

I started my career writing C and C++ code. I always enjoyed designing efficient algorithms. I have experienced many an *Aha!* moments, when I saw how powerful and creative pointer twiddling could be!

For example, reversing a singly linked list using three node pointers is a well known algorithm. We scan the list once and reverse it by changing the pointer fields of each node. The three pointer variables guide the reversal process.

I have come across many such pointer tricks and have used them as needed.

I was next initiated into the world of multi-threading! Variables became shared states between threads! My bagful of tricks was still valid; however, changing state needed a lot of care, to stay away from insidious threading bugs.

The real world is never picture perfect and someone forgot to synchronize a data structure.

Thankfully we started using C++, which had another bagful of tricks, to control the state sharing. You could now make objects immutable!

For example, we were able to implement the readers/writer locking pattern effectively. Immutable objects could be shared without worry among thousands of readers!

We slept easier, the code worked as expected, and all was well with the world!

I soon realized the reason it worked well! Immutability was finally helping us better understand the state changes!

The sands of time kept moving and I discovered functional programming.

I could very well see why writing side-effect free code worked! I was hooked and started playing with Scala, Clojure, and Erlang. Immutability was the norm here.

However, I wondered how the traditional algorithms would look like in a functional setting--and started learning about it.

A data structure is never mutated in place. Instead, a new version of the data structure is created. The strategy of copy on write with maximized sharing was an intriguing one! All that careful synchronization is simply not needed!

The languages come equipped with garbage collection. So, if a version is not needed anymore, the runtime would take care of reclaiming the memory.

All in good time though! Reading this book will help you see that we need not sacrifice algorithmic performance while avoiding in-place mutation!

What this book covers

Chapter 1, *Why Functional Programming?*, takes you on a whirlwind tour of the functional programming (FP) paradigm. We try to highlight the many advantages FP brings to the table when compared with the imperative programming paradigm. We discuss FP's higher level of abstraction, being declarative, and reduced boilerplate. We talk about the problem of reasoning about the state change. We see how being immutable helps realize "an easier to reason about system".

Chapter 2, *Building Blocks*, provides a whirlwind tour of basic concepts in algorithms. We talk about the Big O notation for measuring algorithm efficiency. We discuss the space time trade-off apparent in many algorithms. We next look at referential transparency, a functional programming concept. We will also introduce you to the notion of persistent data structures.

Chapter 3, *Lists*, looks at how lists are implemented in a functional setting. We discuss the concept of persistent data structures in depth here, showing how efficient functional algorithms try to minimize copying and maximize structural sharing.

Chapter 4, *Binary Trees*, discusses binary trees. We look at the traditional binary tree algorithms, and then look at Binary Search Trees.

Chapter 5, *More List Algorithms*, shows how the prepend operation of lists is at the heart of many algorithms. Using lists to represent binary numbers helps us see what lists are good at. We also look at greedy and backtracking algorithms, with lists at the heart.

Chapter 6, *Graph Algorithms*, looks at some common graph algorithms. We look at graph traversal and topological sorting, an important algorithm for ordering dependencies.

Chapter 7, *Random Access Lists*, looks at how we could exploit Binary Search Trees to access a random list element faster.

Chapter 8, *Queues*, looks at First In First Out (FIFO) queues. This is another fundamental data structure. We look at some innovative uses of lists to implement queues.

Chapter 9, *Streams, Laziness, and Algorithms*, looks at lazy evaluation, another FP feature. This is an important building block for upcoming algorithms, so we refresh ourselves with some deferred evaluation concepts.

Chapter 10, *Being Lazy – Queues and Deques*, looks at double-ended queues, which allow insertion and deletion at both ends. We first look at the concept of amortization. We use lazy lists to improve the queue implementation presented earlier, in amortized constant time. We implement deques also using similar techniques.

Chapter 11, *Red-Black Trees*, shows how balancing helps avoid degenerate Binary Search Trees. This is a comparatively complex data structure, so we discuss each algorithm in detail.

Chapter 12, *Binomial Heaps*, covers heap implementation offering very efficient merge operation. We implement this data structure in a functional setting.

Chapter 13, *Sorting*, talks about typical functional sorting algorithms.

What you need for this book

You need to install Scala and Clojure. All the examples were tested with Scala version 2.11.7. The Clojure examples were tested with Clojure version 1.6.0. You don't need any IDE as most of the examples are small enough, so you can key them in the REPL (Read/Eval/Print Loop).

You also need a text editor. Use whichever you are comfortable with.

Who this book is for

The book assumes some familiarity with basic data structures. You should have played with fundamental data structures like linked lists, heaps, and binary trees. It also assumes that you have written some code in a functional language.

Scala is used as an implementation language. We do highlight related Clojure features too. The idea is to illustrate the basic design principles.

We explain the language concepts as needed. However, we just explain the basics and give helpful pointers, so you can learn more by reading the reference links.

We try to site links that offer hands-on code snippets, so you can practice them yourself.

Walking through an algorithm and discussing the implementation line by line is an effective aid to understanding.

A lot of thought has gone into making helpful diagrams. Quizzes and exercises are included, so you can apply what you've learned.

All the code is available online. We strongly advocate keying in the code snippets though, to internalize the principles and techniques.

Welcome to the wonderland of functional data structures and algorithms!

Conventions

In this book, you will find a number of text styles that distinguish between different kinds of information. Here are some examples of these styles and an explanation of their meaning.

Code words in text, database table names, folder names, filenames, file extensions, pathnames, dummy URLs, user input, and Twitter handles are shown as follows: "The following function f has a side effect, though."

A block of code is set as follows:

```
user=> (def v [7 11 19 52 42 72])
#'user/v
user=> (def v1 (conj v 52))
#'user/v1
```

If there is a line (or lines) of code that needs to be highlighted, it is set as follows:

```scala
scala> def pop(queue: Fifo): (Int, Fifo) = {
     |   queue.out match {
     |     case Nil => throw new IllegalArgumentException("Empty queue");
     |     case x :: Nil => (x, queue.copy(out = queue.in.reverse, Nil))
     |     case y :: ys => (y, queue.copy(out = ys))
     |   }
     | }
pop: (queue: Fifo)(Int, Fifo)
```

New terms and **important words** are shown in bold. Words that you see on the screen, for example, in menus or dialog boxes, appear in the text like this: "Clicking the **Next** button moves you to the next screen."

Warnings or important notes appear in a box like this.

Tips and tricks appear like this.

Reader feedback

Feedback from our readers is always welcome. Let us know what you think about this book—what you liked or disliked. Reader feedback is important for us as it helps us develop titles that you will really get the most out of.

To send us general feedback, simply e-mail feedback@packtpub.com, and mention the book's title in the subject of your message.

If there is a topic that you have expertise in and you are interested in either writing or contributing to a book, see our author guide at www.packtpub.com/authors.

Customer support

Now that you are the proud owner of a Packt book, we have a number of things to help you to get the most from your purchase.

Downloading the example code

You can download the example code files for this book from your account at `http://www.p acktpub.com`. If you purchased this book elsewhere, you can visit `http://www.packtpub.c om/support`and register to have the files e-mailed directly to you.

You can download the code files by following these steps:

1. Log in or register to our website using your e-mail address and password.
2. Hover the mouse pointer on the **SUPPORT** tab at the top.
3. Click on **Code Downloads & Errata**.
4. Enter the name of the book in the **Search** box.
5. Select the book for which you're looking to download the code files.
6. Choose from the drop-down menu where you purchased this book from.
7. Click on **Code Download**.

Once the file is downloaded, please make sure that you unzip or extract the folder using the latest version of:

- WinRAR / 7-Zip for Windows
- Zipeg / iZip / UnRarX for Mac
- 7-Zip / PeaZip for Linux

The code bundle for the book is also hosted on GitHub at `https://github.com/PacktPubl ishing/Learning-Functional-Data-Structures-and-Algorithms`. We also have other code bundles from our rich catalog of books and videos available at `https://github.com/P acktPublishing/`. Check them out!

Downloading the color images of this book

We also provide you with a PDF file that has color images of the screenshots/diagrams used in this book. The color images will help you better understand the changes in the output. You can download this file from `https://www.packtpub.com/sites/default/files/down loads/LearningFunctionDataStructuresandAlgorithms_ColorImages.pdf`.

Errata

Although we have taken every care to ensure the accuracy of our content, mistakes do happen. If you find a mistake in one of our books—maybe a mistake in the text or the code—we would be grateful if you could report this to us. By doing so, you can save other readers from frustration and help us improve subsequent versions of this book. If you find any errata, please report them by visiting http://www.packtpub.com/submit-errata, selecting your book, clicking on the **Errata Submission Form link**, and entering the details of your errata. Once your errata are verified, your submission will be accepted and the errata will be uploaded to our website or added to any list of existing errata under the Errata section of that title.

To view the previously submitted errata, go to https://www.packtpub.com/books/content/support and enter the name of the book in the search field. The required information will appear under the Errata section.

Piracy

Piracy of copyrighted material on the Internet is an ongoing problem across all media. At Packt, we take the protection of our copyright and licenses very seriously. If you come across any illegal copies of our works in any form on the Internet, please provide us with the location address or website name immediately so that we can pursue a remedy.

Please contact us at copyright@packtpub.com with a link to the suspected pirated material.

We appreciate your help in protecting our authors and our ability to bring you valuable content.

Questions

If you have a problem with any aspect of this book, you can contact us at questions@packtpub.com, and we will do our best to address the problem.

1
Why Functional Programming?

What is **functional programming** (**FP**)? Why is it talked about so much?

A programming paradigm is a style of programming. FP is a programming paradigm characterized by the *absence of side effects*.

In FP, *functions* are the primary means of structuring code. The FP paradigm advocates using pure functions and stresses on *immutable data structures*. So we don't mutate variables, but pass a state to function parameters. Functional languages give us lazy evaluation and use recursion instead of explicit loops. Functions are first-class citizens like numbers or strings. We pass functions as argument values, just like a numeric or string argument. This ability to pass functions as arguments allows us to compose behavior, that is, cobble together something entirely new from existing functions.

In this chapter, we will take a whirlwind tour of functional programming. We will look at bits of code and images to understand the concepts. This will also lay a nice foundation for the rest of the book. We will use the functional paradigm and see how it changes the way we think about data structures and algorithms.

This chapter starts with a look at the concept of abstraction. We will see why abstractions are important in programming. FP is a *declarative style* of programming, similar to **Structured Query Language** (**SQL**). Because it is declarative, we use it to tell *what* we want the computer to do, rather *how* it should do it. We will also see how this style helps us stay away from writing common, repetitive boilerplate code.

Passing functions as arguments to other, higher order functions is the central idea in FP; we look at this next. We will also see how to stay away from null checks. Controlled state change allows us to better reason our code. Being immutable is the key for creating code that would be easier to reason about.

Next, we will see how recursion helps us realize looping without mutating any variables. We will wrap up the chapter with a look at lazy evaluation, copy-on-write, and functional composition.

The imperative way

We keep contrasting FP with the imperative style of programming. What do we mean by imperative style, though?

The imperative programming style is embodied by a sequence of commands modifying a program's state. A simple example of this is a for loop. Consider the following pseudo code snippet to print all the elements of an array:

```
x = [1,2,3,4...] // an array, x.size tells the number of array elements
for( int i = 0; i < x.size; ++i ) {
        println(x[i])
    }
```

Here is a pictorial rendering of the concepts:

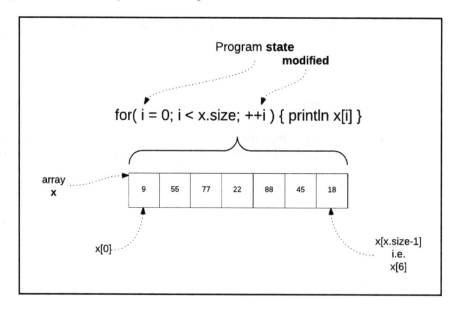

As the figure shows, the for loop establishes an initial state by setting the variable i to 0. The variable is incremented every time the loop is repeated; this is what we mean by the state being modified. We keep reading and modifying the state, that is, the loop variable, until there are no elements left in the array.

FP advocates staying away from any state modification. It gives us tools so we don't worry about *how* to loop over a collection; instead, we focus on *what* we need to do with each element of the collection.

Higher level of abstraction

FP allows us to work at a higher level of abstraction. Abstraction is selective ignorance. The world we know of runs on abstractions. If we say, "Give me sweet condensed milk frozen with embedded dry fruits' someone might really have a hard time understanding it. Instead, just say "ice-cream"! Aha! It just falls into place and everyone is happy.

Abstractions are everywhere. An airplane is an abstraction, so is a sewing machine or a train. When we wish to travel by train, we buy a ticket, board the train, and get off at the destination. We really don't worry about how the train functions. We simply can't be expected to deal with all the intricacies of a train engine. As long as it serves our purpose, we ignore details related to how an engine really works.

What are the benefits of an abstraction? We don't get bogged down into unnecessary details. Instead, we focus on the task at hand by applying higher level programming abstractions.

Compare the preceding for loop with the functional code snippet:

```scala
scala> val x = Array(1,2,3,4,5,6)
x: Array[Int] = Array(1, 2, 3, 4, 5, 6)
scala> x.foreach(println _)
1
2
...
```

We simply focus on the task at hand (print each element of an array) and don't care about the mechanics of a for loop. The functional version is more abstract.

As software engineers, when we implement an algorithm and run it, we are intentionally ignoring many details.

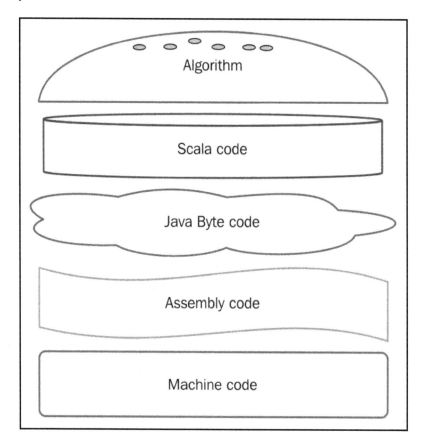

We know that the preceding sandwich stack somehow works and faithfully translates the algorithm into runnable code.

Applying higher level abstractions is commonly done in functional languages. For example, consider the following Clojure REPL snippet:

```
user=> ((juxt (comp str *) (comp str +)) 1 2 3 4 5)
["120" "15"]
```

We are juxtaposing two functions; each of these are in turn composed of the `str` function and an arithmetic function:

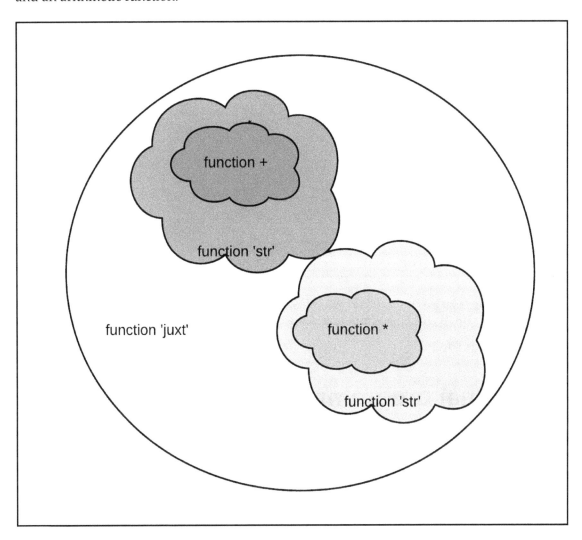

We just don't worry about how it works internally. We just use high-level abstractions cobbled together with existing pieces of *abstract* functionality.

We will be looking at **abstract data types** (**ADT**) closely in this book. For example, when we talk about a stack, we think of the **Last In First Out** (**LIFO**) order of access. However, this ADT is realized by implementing the stack via a data structure, such as a linked list or an array.

Here is a figure showing the **First In First Out** (**FIFO**) ADT in action. The FIFO queue is the normal queue we encounter in life, for example, queuing up at a booking counter. The earlier you get into a queue, the sooner you get serviced.

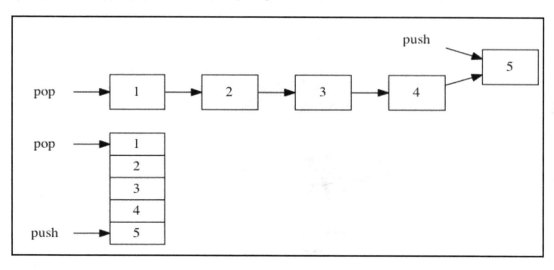

The FIFO queue is an ADT. True that we think of it as an ADT; however, as shown in the preceding diagram, we can also implement the queue backed by either an array or a linked list.

Functional programming is declarative

When we use SQL, we just express our intent. For example, consider this:

```
mysql> select count(*) from book where author like '%wodehouse%';
```

We just say what we are looking for. The actual mechanism that gets the answer is hidden from us. The following is a little too simplistic but suitable example to prove the point.

The SQL engine will have to loop over the table and check whether the author column contains the `wodehouse` string. We really don't need to worry about the search algorithm. The `author` table resides on a disk somewhere. The number of table rows that need to be filtered could easily exceed the available memory. The engine handles all such complexities for us though.

We just *declare* our intent. The following Scala snippet is declarative. It counts the number of even elements in the input list:

```scala
scala> val list = List(1, 2, 3, 4, 5, 6)
list: List[Int] = List(1, 2, 3, 4, 5, 6)

scala> list.count( _ % 2 == 0 )
res0: Int = 3
```

The code uses a higher order function, namely `count`. This takes another function, a predicate, as an argument. The line loops over each list element, invokes the argument predicate function, and returns the `count`.

Here is another example of Clojure code that shows how to generate a combination of values from two lists:

```
user=> (defn fun1 [list1 list2]
  #_=>    (for [x list1 y list2]
  #_=>       (list x y)))
#'user/fun1
user=> (fun1 '(1 2 3) '(4 5 6))
((1 4) (1 5) (1 6) (2 4) (2 5) (2 6) (3 4) (3 5) (3 6))
```

Note the code used to generate the combination. We use *for comprehension* to just state what we need done and it would be done for us.

No boilerplate

Boilerplate code consists sections of the *same code* written again and again. For example, writing loops is boilerplate, as is writing getters and setters for `private` class members.

As the preceding code shows, the loop is implicit:

```scala
scala> List(1, 2, 3, 4, 5) partition(_ % 2 == 0)
res3: (List[Int], List[Int]) = (List(2, 4),List(1, 3, 5))
```

We just wish to separate the odd and even numbers. So we just specify the criteria via a function, an anonymous function in this case. This is shown in the following image:

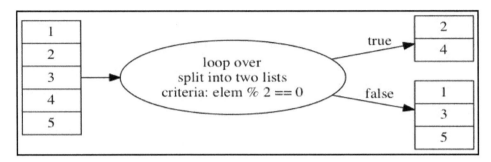

What is boilerplate? It is a for loop, for example. In the imperative world, we code the loop ourselves. We need to tell the system how to iterate over a data structure.

Isn't Scala code just to the point? We tell what we need and the loop is implied for us. No need to write a for loop, no need to invent a name for the loop variable, and so on. We just got rid of the boilerplate.

Here is a Clojure snippet that shows how to multiply each element of a vector by 2:

```clojure
user=> (map * (repeat 2) [1 2 3 4 5])
(2 4 6 8 10)
```

The map function hides the loop from us. Then (repeat 2) function call generates an *infinite sequence*.

So we are just saying this: for the input sequence [1 2 3 4 5], create another lazy sequence of 2's. Then use the `map` function to multiply these two sequences and output the result. The following figure depicts the flow:

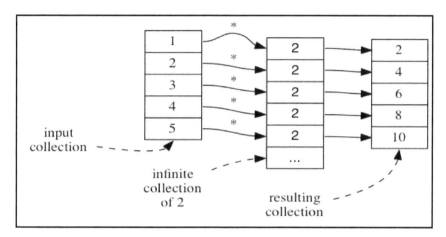

Compare this with an imperative language implementation. We would have needed a loop and a list to collect the result. Instead, we just say what needs to be done.

Higher order functions

Unix shells allow us to express computations as pipelines. Small programs called **filters** are *piped* together in unforeseen ways to ensure they work together. For example, refer to this code:

```
~> seq 100 | paste -s -d '*' | bc
```

This is a very big number (obviously, as we just generated numbers from 1 to 100 and multiplied them together). There is looping involved, of course. We need to generate numbers from 1 to 100, connect them together via `paste`, and pass these on to `bc`. Now consider the following code:

```
scala> val x = (1 to 10).toList
x: List[Int] = List(1, 2, 3, 4, 5, 6, 7, 8, 9, 10)

scala> x.foldLeft(1) { (r,c) => r * c }
res2: Int = 3628800
```

Writing a for loop with a counter and iterating over the elements of the collection one by one is shown in the preceding code. We simply don't have to worry about visiting all the elements now. Instead, we start thinking of how we can process each element.

Here is the equivalent Clojure code:

```
user=> (apply * (range 1 11) )
3628800
```

The following figure shows how the code works:

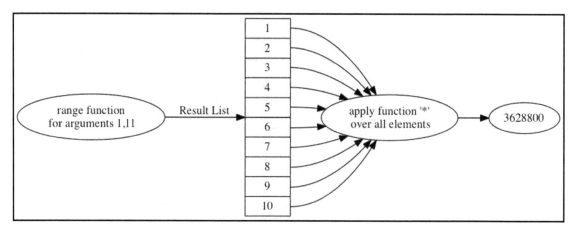

Scala's `foldLeft` and Clojure's `apply` are higher order functions. They help us avoid writing boilerplate code. The ability to supply a function brings a lot of flexibility to the table.

Eschewing null checks

Higher order functions help us succinctly express logic. There is an alternative paradigm that expresses logic without resorting to null checks. Refer to `http://www.infoq.com/presentations/Null-References-The-Billion-Dollar-Mistake-T ony-Hoare` on why nulls are a bad idea; it comes directly from its inventor.

Here is a Scala collection with some strings and numbers thrown in; it is written using the alternative paradigm:

```
scala> val funnyBag = List("1", "2", "three", "4", "one hundred seventy
five")
funnyBag: List[String] = List(1, 2, three, 4, one hundred seventy five)
```

We want to extract the numbers out of this collection and sum them up:

```
scala> def toInt(in: String): Option[Int] = {
     |     try {
     |        Some(Integer.parseInt(in.trim))
     |     } catch {
     |         case e: Exception => None
     |     }
     | }
toInt: (in: String)Option[Int]
```

We return an `Option` container as a return value. If the string was successfully parsed as a number, we return `Some[Int]`, holding the converted number.

The following diagram shows the execution flow:

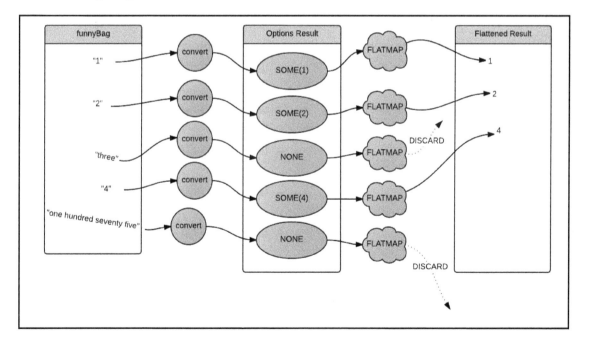

In the case of a failure, we return None. Now the following higher order function skips None and just flattens Some. The flattening operation pulls out the result value out of Some:

```scala
scala> funnyBag.flatMap(toInt)
res1: List[Int] = List(1, 2, 4)
```

We can now do further processing on the resulting number list, for example, summing it up:

```scala
scala> funnyBag.flatMap(toInt).sum
res2: Int = 7
```

Controlling state changes

Let me share a debugging nightmare with you. We had a multithreaded system written in C++. The state was carefully shared, with concurrent access protected by explicit mutex locks. A team member–ugh–forgot to acquire a lock on a shared data structure and all hell broke loose.

The team member was a senior programmer; he knew what he was doing. He just forgot the locking. It took us some nights full of stack trace to figure out what the issue was.

Writing concurrent programs using shared memory communication can very easily go wrong.

In the book *Java Concurrency in Practice*, the authors show us how easy it is for *internal mutable state* to escape (http://jcip.net/). Tools, such as Eclipse, make it easy to generate getters, and before you know, a reference escapes and all hell could break loose.

The encapsulation is fine. The mutable state, in this case an array, could be inadvertently exposed. For example, using a getter method, the array reference can be obtained by the outside world. Two or more threads could then try mutating it and everything goes for a toss:

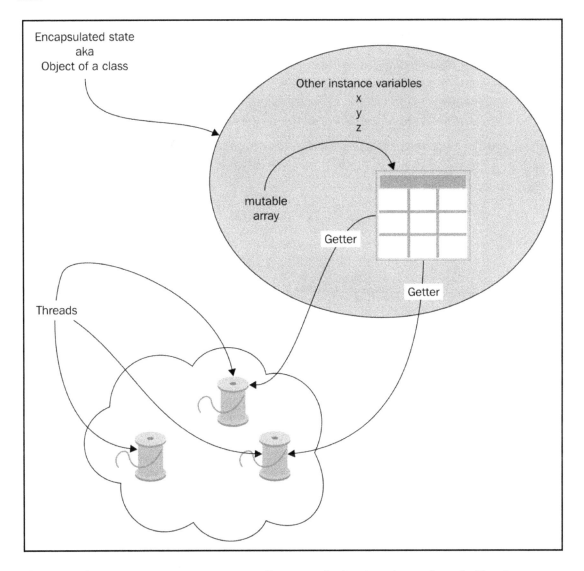

We cannot ignore concurrency anymore. Program design is going to be ruled by the machine design; having a multicore machine is the norm.

It is too hard to make sure the state changes are controlled. If instead, we know that a data structure does not change once it is created, reasoning becomes far easier. There is no need to acquire/release locks as a shared state never changes.

Recursion aids immutability

Instead of writing a loop using a mutable loop variable, functional languages advocate recursion as an alternative. Recursion is a widely used technique in imperative programming languages, too. For example, quicksort and binary tree traversal algorithms are expressed recursively. Divide and conquer algorithms naturally translate into recursion.

When we start writing recursive code, we don't need mutable loop variables:

```
scala> import scala.annotation.tailrec
import scala.annotation.tailrec
scala> def factorial(k: Int): Int = {
     |     @tailrec
     |     def fact(n: Int, acc: Int): Int = n match {
     |       case 1 => acc
     |       case _ => fact(n-1, n*acc)
     |     }
     |     fact(k, 1)
     | }
factorial: (k: Int)Int

scala> factorial(5)
res0: Int = 120
```

Note the `@tailrec` annotation. Scala gives us an option to ensure that **tail call optimization (TCO)** is applied. TCO rewrites a recursive tail call as a loop. So in reality, no stack frames are used; this eliminates the possibility of a stack overflow error.

Here is the equivalent Clojure code:

```
user=> (defn factorial [n]
  #_=>    (loop [cur n fac 1]
  #_=>      (if (= cur 1)
  #_=>        fac
  #_=>          (recur (dec cur) (* fac cur) ))))
#'user/factorial
user=> (factorial 5)
120
```

The following diagram shows how recursive calls use stack frames:

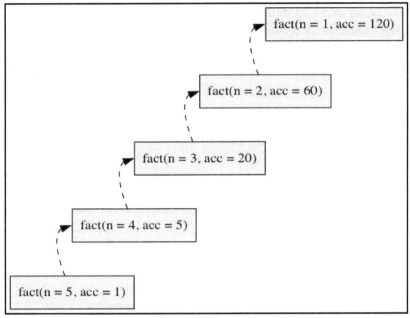

Clojure's special form, *recur*, ensures that the TCO kicks in.

Note how these versions are starkly different than the one we would write in an imperative paradigm.

Instead of explicit looping, we use recursion so we wouldn't need to change any state, that is, we wouldn't need any mutable variables; this aids immutability.

Copy-on-write

What if we have never mutated data? When we need to update, we could copy and update the data. Consider the following Scala snippet:

```
scala> val x = 1 to 10
x: scala.collection.immutable.Range.Inclusive = Range(1, 2, 3, 4, 5, 6, 7,
8, 9, 10)

scala> x  map (_ / 2) filter ( _ > 0 ) partition ( _ < 2 )
res4: (scala.collection.immutable.IndexedSeq[Int],
scala.collection.immutable.IndexedSeq[Int]) = (Vector(1, 1),Vector(2, 2, 3,
3, 4, 4, 5))
```

Here is a figure showing all of the copying in action:

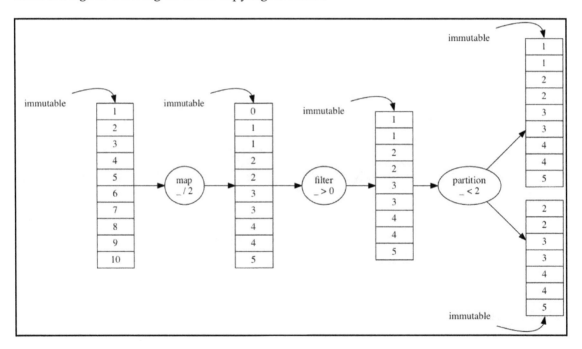

This is *copy-on-write* semantics: we make a new data structure every time a change happens to the original data structure. Note the way the filter works. Just one element is removed–the first one–but we cannot simply remove the element from the input vector itself and pass it on to the partition logic.

This solves the problem of the leaking getter. As data structures are immutable, they could be freely shared among different threads of execution. The state is still shared, but just for reading!

What happens when the input is too large? It would end up in a situation where too much of data is copied, wouldn't it? Not really! There is a behind-the-scenes process called structural sharing. So most of the copying is avoided; however, immutability semantics are still preserved. We will be looking at this feature closely when we study Chapter 3, *Lists* .

Laziness and deferred execution

To deal with excessive copying, we can resort to a feature called **deferred processing,** also known as, *lazy collections*. A collection is lazy when all of its elements are not realized at the time of creation. Instead, elements are computed on demand.

Let's write a program to generate numbers from 1 to 100. We wish to check which numbers are evenly divisible by 2, 3, 4, and 5.

Let's generate a lazy collection of the input numbers:

```
scala> val list = (1 to 100).toList.view
list: scala.collection.SeqView[Int,List[Int]] = SeqView(...)
```

We convert an existing Scala collection into a lazy one by calling the view function. Note that the list elements are not printed out, as these are not yet computed.

The following snippet shows a very simple predicate method that checks whether the number n is evenly divisible by d:

```
scala> def isDivisibleBy(d: Int)(n: Int) = {
     |    println(s"Checking ${n} by ${d}")
     |    n % d == 0
     | }
isDivisibleBy: (d: Int)(n: Int)Boolean
```

We write a method `isDivisibleBy` in the *curried form*. We have written the `isDivisibleBy` as a *series* of functions, each function taking one argument. In our case, n is 2. We do this so we can partially apply functions to the divisor argument. This form helps us easily generate functions for divisors 2, 3, 4, and 5:

```scala
scala> val by2 = isDivisibleBy(2) _
by2: Int => Boolean = <function1>

scala> val by3 = isDivisibleBy(3) _
by3: Int => Boolean = <function1>

scala> val by4 = isDivisibleBy(4) _
by4: Int => Boolean = <function1>

scala> val by5 = isDivisibleBy(5) _
by5: Int => Boolean = <function1>
```

We can test the preceding functions by entering the code on the REPL, as shown here:

```scala
scala> by3(9)
Checking 9 by 3
res2: Boolean = true

scala> by4(11)
Checking 11 by 4
res3: Boolean = false
```

Now we write our checker:

```scala
scala> val result = list filter by2 filter by3 filter by4 filter by5
result: scala.collection.SeqView[Int,List[Int]] = SeqViewFFFF(...)
scala> result.force
Checking 1 by 2
Checking 2 by 2
Checking 2 by 3
Checking 3 by 2
Checking 4 by 2
...
Checking 60 by 2
Checking 60 by 3
Checking 60 by 4
Checking 60 by 5
...
res1: List[Int] = List(60)
```

Note that when 2 is checked by 2 and okayed, it is checked by 3. All the checks happen at the same time and the copying is elided.

Note the `force` method; this is the opposite of the `view` method. The `force` method converts the collection back into a strict one. For a strict collection, all the elements are processed. Once the processing is done, a collection with just the number 60 is returned.

Composing functions

We programmers, love reusing code. We use libraries and frameworks, so we use something that already exists out there. No one wants to reinvent the same wheel again. For example, here is how we could weed out zeroes from a Clojure vector:

```
user=> (filter (complement zero?) [0 1 2 0 3 4])
(1 2 3 4)
```

The following diagram shows the functional composition:

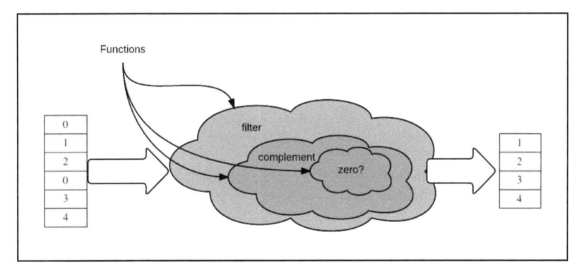

We *composed* behavior by composing predicate functions; we did this using `complement` to negate the `zero?` predicate function. The `zero?` predicate returns `true` if its input is 0; if not, it returns `false`.

Given a list of numbers, the following Scala snippet checks whether the sequence is strictly increasing:

```
scala> val x = List(1, 2, 3, 4, 5, 6, 7)
x: List[Int] = List(1, 2, 3, 4, 5, 6, 7)

scala> val y = x zip x.drop(1)
y: List[(Int, Int)] = List((1,2), (2,3), (3,4), (4,5), (5,6), (6,7))

scala> y forall (x => x._1 < x._2)
res2: Boolean = true
```

Just imagine how we would do this in an imperative language.

Using `zip`, we get each number and its successor as a pair. We pass in a function to know whether the first element of the pair is less than the second.

Here goes the Clojure implementation. First we define a function that takes a list of two elements, de-structures it into its elements, and checks whether these two elements are strictly increasing:

```
user=> (defn check? [list]
  #_=>    (let [[x y] list]
  #_=>      (> y x)))
```

Here is a quick test:

```
user=> (check? '(21 2))
false
user=> (check? '(1 2))
true
```

 Note that the check? function is a *pure function*. It works just on its input and nothing else.

Next comes the pair generation; here, each element is paired with its successor:

```
user=> (defn gen-pairs [list]
  #_=>    (let [x list
  #_=>          y (rest list)]
  #_=>      (partition 2 (interleave x y))))
```

Testing it gives the following:

```
user=> (gen-pairs '(1 2 3 4))
((1 2) (2 3) (3 4))
```

Now comes the magic! We compose these two functions to check whether the input is strictly increasing:

```
user=> (defn strictly-increasing? [list]
  #_=>    (every? check? (gen-pairs list)))
#'user/strictly-increasing?
```

Testing it gives the following:

```
user=> (strictly-increasing? '(1 2 3 4))
true
user=> (strictly-increasing? '(1 2 3 4 2))
false
```

See how succinct it becomes when we start composing functions:

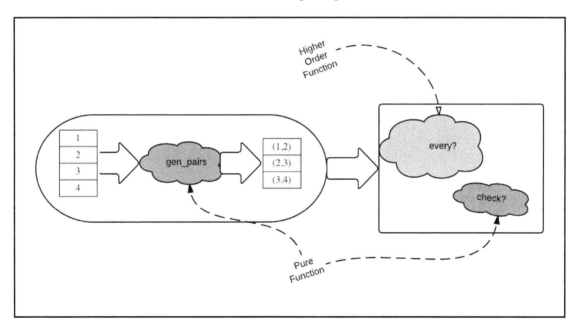

Note that the functions check? and every? are *pure functions*. The check? function predicate works on the input only. The gen_pairs is also a pure function. It is, in turn, composed of the interleave and partition functions.

The `every?` function is also a higher order function. We never wrote any loops (not even any recursion). We composed behavior out of existing pieces. The fun thing is the pieces don't need to know about the composition. We just combine independent processing pieces together.

For a great treatment of the advantages FP brings to the table, we wholeheartedly recommend *Functional Thinking– Paradigm Over Syntax* (`http://shop.oreilly.com/product/0636920029687.do`).

Summary

We looked at some prominent reasons to adopt the FP paradigm.

Firstly, we saw what is *imperative programming* and the notion of state modification.

FP allows us to program at a higher level of abstraction. We looked at some common examples of applying such abstractions.

We saw how FP encourages us to compose systems from existing building blocks. These blocks themselves, in turn, could have been composed out of other smaller blocks. This is an incredibly powerful way to reuse code.

The declarative style of programming is easily seen in how SQL queries work. This allows us to work at a higher level of abstraction.

FP promotes this same declarative style. For example, we normally use implied loops. Implied loops in FP are similar to how Unix shell filters process data.

Controlling changes to a program's state is way too hard. We saw how important this is, given the multithreaded world we developers live in. We saw how FP makes it a breeze by dealing with mostly immutable data structures.

In the next chapter, we will look at some fundamental concepts in data structures and algorithms.

2
Building Blocks

This chapter serves as a refresher on some fundamentals concepts.

How fast could an algorithm run? How does it fare when you have ten input elements versus a million? To answer such questions, we need to be aware of the notion of algorithmic complexity, which is expressed using the *Big O* notation. An *O(1)* algorithm is faster than $O(log_n)$, for example.

What is this notation? It talks about measuring the efficiency of an algorithm, which is proportional to the number of data items, N, being processed.

This chapter starts with a look at the O notation. Space/time trade-off is another important aspect of algorithm design. Let's look at a dynamic programming problem to better understand this fundamental notion.

Next, we will look at vectors and list data structures and note the trade-offs.

We will conclude by looking at the complexities of some functional idioms.

By the end of this chapter, you will have a good understanding of algorithmic complexities. These concepts are essential for understanding the rest of the book.

The Big O notation

In simple words, this notation is used to describe how fast an algorithm will run. It describes the *growth* of the algorithm's running time versus the size of input data.

Here is a simple example. Consider the following Scala snippet, reversing a linked list:

```scala
scala> def revList(list: List[Int]): List[Int] = list match {
     |    case x :: xs => revList(xs) ++ List(x)
     |    case Nil => Nil
     | }
revList: (list: List[Int])List[Int]
scala> revList(List(1,2,3,4,5,6))
res0: List[Int] = List(6, 5, 4, 3, 2, 1)
```

A quick question for you: how many times does the first case clause, namely `case x ::
xs => revList(xs) ++ List(x)`, match a list of six elements? Note that the clause
matches when the list is *non-empty*. When it matches, we reduce the list by one element and
recursively call the method.

It is easy to see the clause matches six times. As a result, the list method, ++, also gets
invoked four times. The ++ method takes time and is directly proportional to the *length of
the list on left-hand side*. (In the next chapter, we will look at this operation in detail.)

Here is a plot of the number of iterations against time:

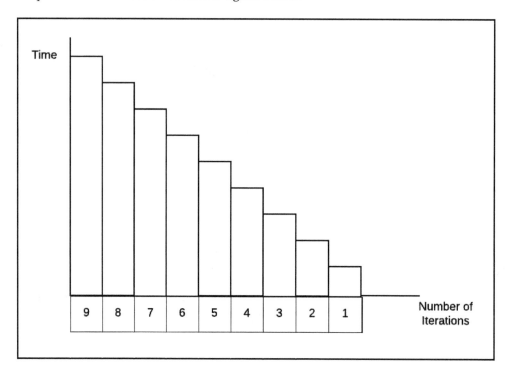

To reverse a list of nine elements, we iterate over the elements 55 times (**9+8+7+6+5+4+3+2+1**). For a list with *20* elements, the number of iterations is *210*.

Here is a table showing some more example values:

Num Elems	Num Iterations	(Num*Num)/2
9	55	40
20	210	200
30	465	450
100	5050	5000
1000	500500	500000

The number of iterations are proportional to $n^2/2$. It gives us an idea of how the algorithm runtime grows for a given number of elements.

The moment we go from 100 to 1,000 elements, the algorithm needs to do 500 times more work.

Another example of a quadratic runtime algorithm is a selection sorting. It keeps finding the minimum element and increases the sorted sublist. The algorithm keeps scanning the list for the next minimum element and hence always has $O(n^2)$ complexity; this is because it does $O(n^2)$ comparisons. Refer to `http://www.studytonight.com/data-structures/selection-sorting` for more information.

Binary search is a very popular search algorithm with a complexity of $O(log_n)$. The succeeding figure shows the growth table for $O(log_n)$.

When the number of input elements jumps from **256** to **10,48,576**, the growth jumps from **8** to **20**. Binary search is a *blazing fast* algorithm as the runtime grows marginally when the number of input elements jump from a couple hundreds to hundred thousand.

Num Elems	logN
256	8
4096	12
16384	14
65536	16
1048576	20

Refer to `https://rob-bell.net/2009/06/a-beginners-guide-to-big-o-notation/` for an excellent description of the O notation.

Refer to the following link that has a graphical representation of the various growth functions:
`https://therecyclebin.files.wordpress.com/2008/05/time-complexity.png`

Space/time trade-off

A trade-off is a balancing act: when we take something, we give away another thing!

Algorithm designs too, at times, trade-off some amount of memory to save on the overall time. Let's look at two problems to better appreciate this important concept.

A word frequency counter

Let's say we have a list of words. The task is to find how many times a word occurs in the list in order to compute every word's *frequency*.

Here is a brute force approach:

```
w <- each word in the list, count <- 1
  w1 <- all other words in the list
    If (w == w1)
        Increment count
            println(w, " = ", count)
```

The following diagram shows the comparisons for the first two elements:

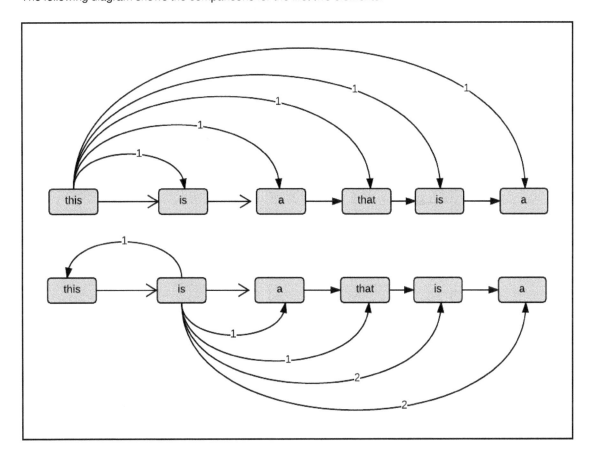

The preceding diagram shows how the algorithm works for the first two words. Each word ends up being compared with other words. Note that even if we know the answer for the word "**is**," we end up recomputing it again.

The algorithm performs $O(n^2)$ comparison. Thus, the runtime complexity of this algorithm is proportional to $O(n^2)$. As the number of items in the list n grows, the time to execute the algorithm grows by its square.

Here is another algorithm that trades off some space to save time. We use an auxiliary data structure, a *hash table*, where the *key* is a word and the *value* the frequency count.

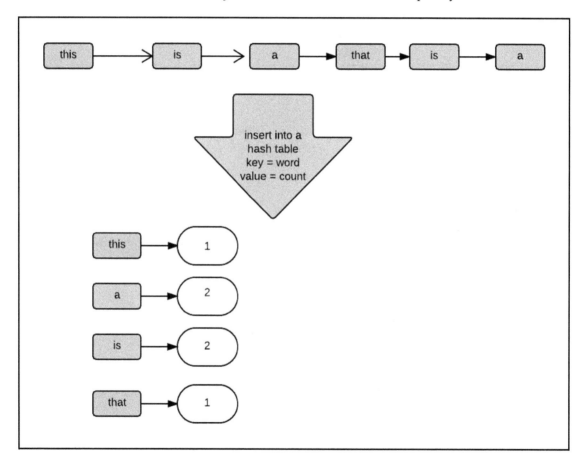

It is a very simple algorithm:

```
For each word w in the input list
    key <- w
```

```
        if (key is not present in the hash table) {
                insert (key, 1) into table
        } else {
          update count for key to count+1
          }
     Iterate over the hash table, and print each (key, count) pair.
```

What is the hash table operation complexity for inserting/updating the pair? It is *effectively constant time*, that is, *O(1)*. So the overall complexity is *O(n)*. Here is a Scala implementation of the algorithm:

```
scala> val words = List("this", "is", "a", "that", "is", "a")
words: List[String] = List(this, is, a, that, is, a)

scala> val freq = words groupBy(x => x) map { y => (y._1, y._2.length) }
freq: scala.collection.immutable.Map[String,Int] = Map(is -> 2, that -> 1,
a -> 2, this -> 1)

scala> freq foreach println
(is,2)
(that,1)
(a,2)
(this,1)
```

Note that we need *O(n)* auxiliary space to store the keys and the associated counts, though.

Exercise for you: Implement the algorithm in Clojure.

Matching string subsets

Let's look at another problem, which again highlights the space time trade-off:

- **Input**: A string s and a dictionary of words, represented by dict
- **Output**: Break the string into those words that are provided in the given dictionary.

For example, if s = "kidsoftheworldunite" and a dictionary {"kids", "the", "unite", "of" "world"} are given, we should return "kids of the world unite".

On the other hand, if s = "allkidsoftheworldunite" is given, then we cannot break it up as there is no all in the dictionary.

Let's look at the algorithm to see how we could arrive at a good solution.

To focus on the core algorithm, we assume the following:

- A dictionary implementation is assumed to be available. We don't need to implement it.
- The dictionary supports only *exact* string lookup.
- The dictionary fits into the memory.

Here is a pseudo code of the first cut, illustrating *recursive backtracking*:

```
break_into_words(s, dictionary) {
    if (dictionary.contains(s))
        return input;
    len = s.len;
    for (i = 1; i < len; i++) {
        prefix = s.substring(0, i);
        suffix = s.substring(i, len);
        if (dictionary.contains(prefix)) {
            words = break_into_words(suffix, dictionary)
            if (words != nil) {
                return prefix + " " + words;
            }
        }
    }
    return nil;
}
```

We try to find the word in the dictionary, and when we meet with failure, we *backtrack* and try another path.

What is the complexity of this algorithm? Let's say we have a dictionary {"x", "xx", "xxx", "xxxx"} and our input string is "xxxxxxxxxy".

Enumerating each and every path involves redoing the checks again and again. We check and fail each time and then enumerate the failure path again. Here is a diagram to better understand the process:

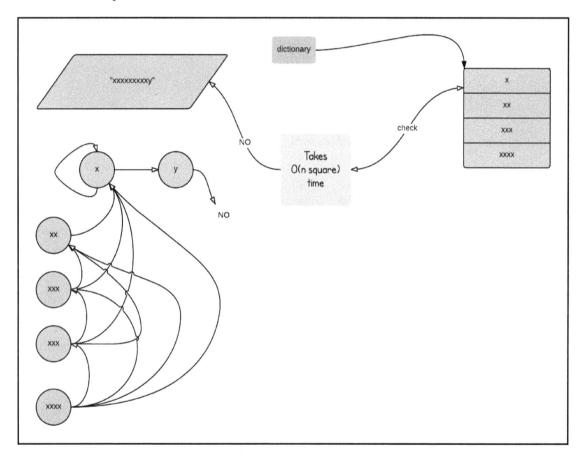

The complexity of the algorithm for this degenerate input is $O(n^2)$.

Do you remember *memoize*, also known as *memorize*, which is the already enumerated path? We assume an associative array, namely ADT (for example, a map), to index by strings and return a string:

```
rememberSuff = ... // Some associative array ADT
break_into_words(s, dictionary) {
        if (dictionary.contains(s))
            return input;
        if (rememberSuff.contains(s) {
            return rememberSuff.get(s); // could be nil
                                        // or a valid suffix
        }
        len = s.len;
        for (I = 1; i < len; i++) {
            prefix = s.substring(0, i);
            suffix = s.substring(i, len);
            if (dictionary.contains(prefix)) {
                words = break_into_words(suffix, dictionary)
                if (words != nil) {
                    rememberSuff.put(s, prefix + " " + words);
                    return prefix + " " + words;
                }
            }
        }
        rememberSuff.put(s, nil); // remember failure
        return nil;
}
```

This version *trades off* some memory for saving the CPU cycles.

We remember the prefix result and reuse it when needed. In other words, the prefix results are *cached*, and at the cost of some memory, we speed up the algorithm.

This is a problem amenable to dynamic programming. When we already have an answer to a subproblem, we use it instead of computing it again.

Storing such answers need memory; at the cost of this, we avoid wasting CPU cycles.

See `https://www.hackerearth.com/notes/dynamic-programming-i-1/` for a nice discussion of this algorithm design technique.

Referential transparency

We appreciate the virtues of caching, but we do this to look at referential transparency, a cornerstone of functional programming.

In the preceding example, note that we are able to cache the results, as the results of the computation are not going to change for the same input. We need not repeat the computations; instead, we could compute the answer once and *save and substitute* it.

In the FP world, where we can substitute a function by its value, the function is called **referentially transparent**. Just like we avoid repeated calls in the previous algorithm, repeated calls to such functions could be avoided by caching the result.

Mathematical functions are referentially transparent. For example, the following Clojure functions are referentially transparent:

```
user=> (* 3 4)
12
user=> (apply + [1 2 3 4 5 6])
21
```

You will always get 21 when you add up 1,2,3,4,5, and 6. Multiplying 3 and 4 will always give the result 12.

 Note that the preceding functions are *pure functions*, as the result depends on the *input arguments* only.

On the other hand, note that the following function f is not referentially transparent:

```
user=> (defn f []
  #_=>   (* 3 (rand)))
user=> (f)
0.20654893720713086
user=> (f)
0.8714300390452085
```

Consider we have the following:

```
user=> (def v (f))
''user/v
user=> v
1.4792154527750396
```

We cannot substitute v with a call of f:

```
user=> (println v v v)
1.4792154527750396 1.4792154527750396 1.4792154527750396
user=> (println (f) (f) (f))
1.6331448272316837 0.8260298442547525 0.9463702072853114
```

Both the results are different.

Pure functions do not have side effects. The following function f has a side effect, though. It returns 9 but prints a message, which is a side effect:

```
user=> (defn f []
  #_=>    (do
  #_=>       (println""Hello worl"")
  #_=>       (identity 9)
  #_=>    )
  #_=> )
```

We can cache the result of a pure function, as shown in the following snippet:

```
user=> (defn slow-func
  #_=>    "This function likes to take a nap first!"
  #_=>    [arg]
  #_=>    (Thread/sleep 2000)
  #_=>    (* 3 arg))
#'user/slow-func
user=> (slow-func 3)
Slow, sleepy execution
9
```

You can call it once and cache-in other words, memorize-the result:

```
user=> (def memoizeIt (memoize slow-func))
#'user/memoizeIt
user=> (memoizeIt 3)
slow, sleepy first execution
9
user=> (memoizeIt 3)
9 // fast
user=> (memoizeIt 3)
9 // fast
```

After you memoize it, the first call will be slow and sleepy. The second and subsequent calls will be blazing fast!

Vectors versus lists

Prepending an element to a linked list is very fast. In fact, it is an *O(1)* operation, meaning the original list is simply pointed at by the new element node. The change happens only at the head of the list. As we don't need to traverse the list at all, this is a fixed cost, that is, *O(1)* operation.

Accessing an element at some index *n* is slower, meaning it is proportional to the number of elements in the list. We need to start at the head and keep traversing the nodes and keep counting. We do this until we reach the nth node. If we access the second last node, we will have traversed almost all of the list.

When any operation could make us look at almost all the elements, the complexity would be *O(n)*. This means it would be proportional to the number of elements.

On the other hand, appending an element to a list is costly when we need to preserve the original list. In the next chapter, we need to *traverse and copy* all the elements, so we will preserve the current version.

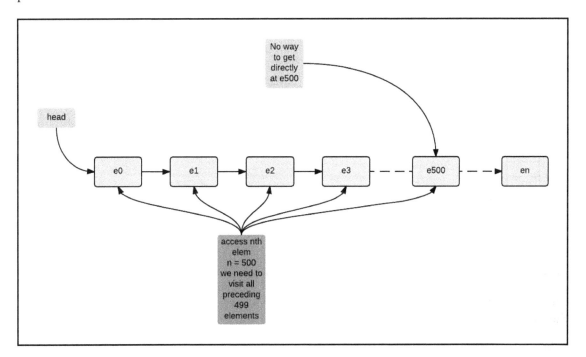

The list data structure does not provide us with *random access*. We will now look at *vectors*, which addresses most of these problems, giving us good performance when we prepend, append, insert in the middle and perform random access operations.

So, how come vectors achieve all of this? It sounds almost magical. The design uses a tree structure internally to localize the copy operation. Here is a conceptual representation of a persistent vector:

The vector internally uses a *trie* structure composed of *internal* and *leaf* nodes. The internal nodes don't hold any data, just pointers to children. On the other hand, the leaf nodes hold just data.

Here is a diagram that shows a vector realized using *trie*:

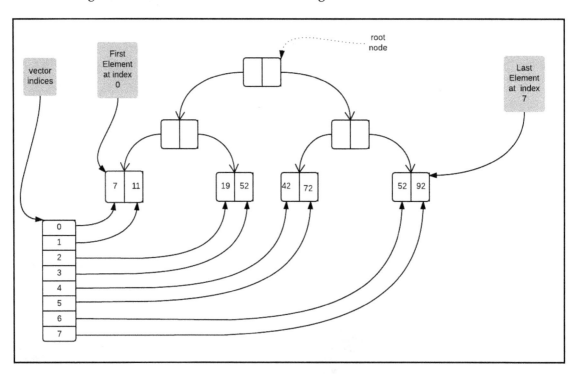

This represents the vector: [7,11,19,52,42,72,52,92]. All the elements are the leaves of the tree.

When we need an element at, say, index 3, we start at the root, go left, and then right to land at the fourth element, namely 52. As mentioned already, we will look at the algorithm in detail in an upcoming chapter.

The internal nodes have two children, also called **branching factors**. Let's look at how this data structure improves a commonly used operation:

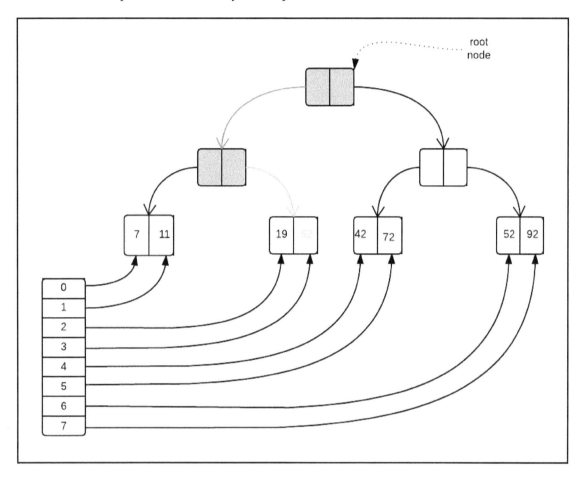

What benefits does this design bring to us? Let's consider the case of removing the last element: 92.

We structurally share all the nodes, except the rightmost one. All the paths from the root to the rightmost node need to be copied.

Note that the changes are pretty localized. We need to copy the last leaf node array and then wipe out the element with the value 92, as shown in the succeeding image.

Let's see the complexity of the append operation. Here is the code snippet, the effect of which we will look at later. The code creates a vector and then appends a new element, number 92, via `conj`:

```
user=> (def v [7 11 19 52 42 72 52])
#'user/v
user=> (def v1 (conj v 92))
#'user/v1
user=> v
[7 11 19 52 42 72 52]
user=> v1
[7 11 19 52 42 72 52 92]
user=>
```

The sharing and copying of nodes is shown in the diagram as shown:

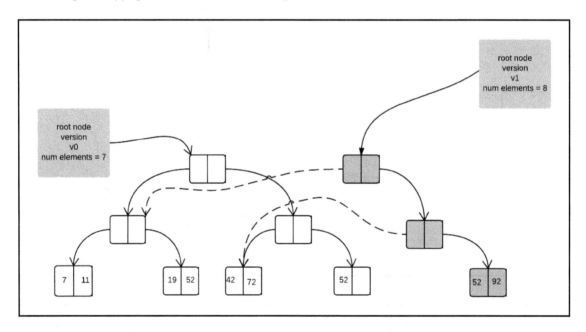

The first six elements (7, 11, 19, 52, 42, 72) are structurally shared; the sharing is shown with a dotted line. Only the last leaf node is copied and the second element is set to 92.

This is a pretty localized operation. It does not depend on the number of elements in the vector. In particular, the cost of appending to a vector with 100 elements or 100000 elements will almost be the same. We just copy the affected path and share everything else.

Updating an element

How would updating an existing vector element look like? In particular, what would happen with the following code:

```
user=> v
[7 11 19 52 42 72 52]
user=> (def v1 (assoc v 5 100))
#'user/v1
user=> v1
[7 11 19 52 42 100 52]
user=>
```

We change the element at the 5th index to 100:

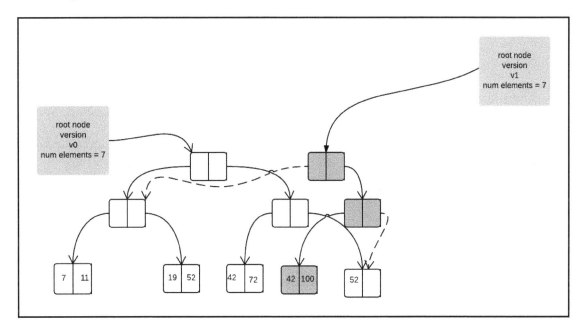

Most of the structure nodes are shared. Copying is done only for the affected path, the one where the update happens.

We copy the leaf node corresponding to the indices 4 and 5 and set the value at the 5th index to 100.

Again, as we can see, the changes are pretty localized and hence the complexity does not depend on the *total number of elements*. Even if the number of elements explode from `100` to `100000`, the operation will be almost as fast.

Not enough nodes

One case is when we don't have enough leaf nodes, for example, when we append the 9th element to the vector, as follows:

```
user=> (def v [7 11 19 52 42 72])
#'user/v
user=> (def v1 (conj v 52))
#'user/v1
```

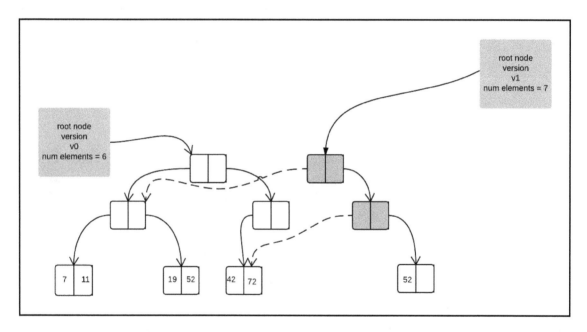

In this case, we grow the tree by adding a leaf node. Again, as we see, there is a controlled change made to grow the tree.

Hopefully, you got a good sense of how the trie structure helps to make a bit of change and is independent of the number of elements in the vector. The complexity of these operations is *effectively constant time*.

Here's an exercise for you: Draw a diagram to show how deleting an element would work, keeping persistence in mind.

For example, it would be instructive to delete the element 52 that was just added. Draw all versions of the internal trie.

Complexities and collections

Let's look at collections and see how complexity helps us to see how they will perform in different situations. We will look at commonly used collections and idioms.

The sliding window

The sliding method allows us to create a sliding window. Here's an example:

```
scala> val list = List(1,2,3,4,5,6)
list: List[Int] = List(1, 2, 3, 4, 5, 6)

scala> val list1 = list.sliding(2,1).toList
list1: List[List[Int]] = List(List(1, 2), List(2, 3), List(3, 4), List(4, 5), List(5, 6))
```

This creates `List[Int]`. Each element of `List` contains the current element and its successor in the original list.

Here is the equivalent code in Clojure:

```
user=> (partition 2 1 '(1 2 3 4 5 6))
((1 2) (2 3) (3 4) (4 5) (5 6))
```

It gives us similar results:

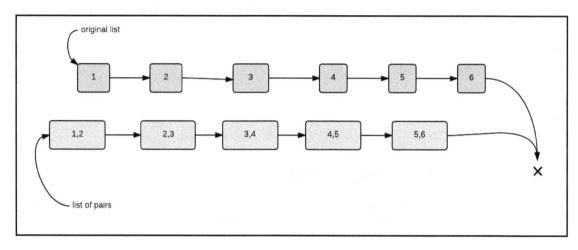

Let's see the complexity of this sliding window code. The *time complexity* is $O(n)$. We need to visit each element of the original list twice: once when it is the first element, and second, when it is the second element. This is clearly proportional to the number of elements in the list, hence $O(n)$.

The *space complexity* is $O(n)$ too. We need space that is proportional to the number of input elements. However, this needs an auxiliary $O(n)$ space too, as we need to store the current element and the next element.

Sliding windows give us a very easy way to look at an element and the next together.

Maps

We have seen maps in action to get a very fast word-counting algorithm.

The groupBy method gives us a way to group elements of a collection. The grouping criterion is decided by the argument function.

For example, here is Scala's groupBy in action:

```scala
scala> List("hello", "world", "scala", "has", "arrived") groupBy { x =>
x.length }
res10: scala.collection.immutable.Map[Int,List[String]] = Map(5 ->
List(hello, world, scala), 7 -> List(arrived), 3 -> List(has))
```

Here is the Clojure version:

```
user=> (group-by count '("hello", "world", "scala", "has", "arrived") )
{5 ["hello" "world" "scala"], 3 ["has"], 7 ["arrived"]}
```

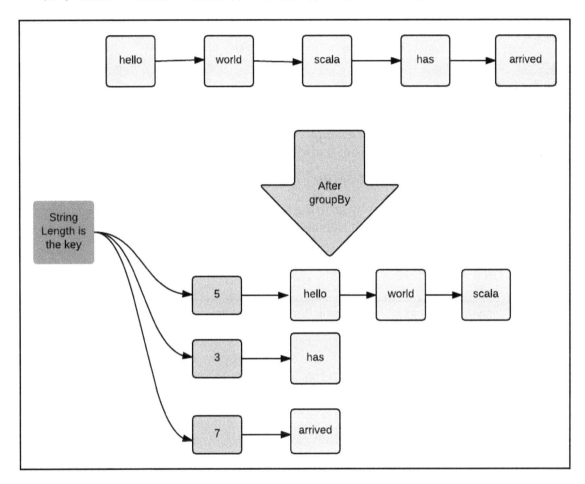

The function groups the string elements by length and collects all the length strings that are equal in the same group. The key is the length of the string, and the value is a list of all the strings that have that length.

The complexity of the groupBy operation involves scanning the input tuple list, which is *O(n)*. We need to visit each tuple element and see which group it should go into.

What is the complexity of populating the map? A map is an associative data structure where we use keys to access values. The underlying implementation is a *Hash Trie*. We will take a detailed look at this data structure in a later chapter. Refer to `http://www.scala-lang.org/docu/files/collections-api/collections_19.html` and `http://www.scala-lang.org/api/current/index.html#scala.collection.immutable.HashMap` for more.

Persistent stacks

To get the **Last In First Out** (**LIFO**) behavior, we use a *stack*. This is a very commonly used data structure in programming. For example, when a method calls another method, which in turn calls another method, a stack of frames is created.

This frame helps us return control from the callee to the caller.

We push elements to a stack, thereby growing them. We pop elements off the stack, thereby shrinking them. It is an error to push elements to a full stack or pop elements off from an empty one.

Here is the stack in action:

```
scala> import scala.collection.immutable.Stack
import scala.collection.immutable.Stack
scala> val s = Stack.empty
s: scala.collection.immutable.Stack[Nothing] = Stack()
```

Here's is an immutable stack:

```
scala> val s1 = s.push(1)
s1: scala.collection.immutable.Stack[Int] = Stack(1)
scala> s
res0: scala.collection.immutable.Stack[Nothing] = Stack()

scala> s1
res1: scala.collection.immutable.Stack[Int] = Stack(1)

scala> val s2 = s1.push(2)
s2: scala.collection.immutable.Stack[Int] = Stack(2, 1)

scala> s2.top
res2: Int = 2

scala> s1.top
res3: Int = 1
```

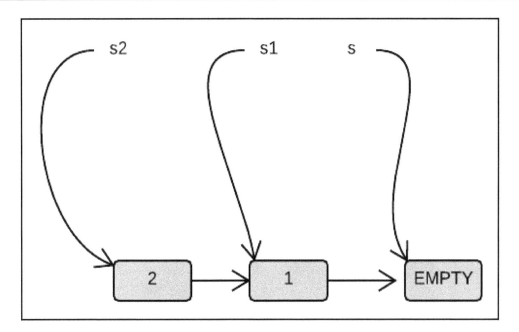

We create three versions: s is the empty stack (which stays empty) and then we have s1 and s2. Alert readers will quickly see this is nothing different from list manipulation:

```
scala> val l = List()
l: List[Nothing] = List()

scala> val l1 = 1 :: l
l1: List[Int] = List(1)

scala> val l2 = 2 :: l1
l2: List[Int] = List(2, 1)

scala> l2.head
res4: Int = 2

scala> l2.tail
res5: List[Int] = List(1)
```

The tail operation equals a popup action and the cons operation is equivalent to a push. The complexity of these operations is *O(1)*.

Persistent FIFO queues

Queues are **First In First Out** (**FIFO**) data structures. A good example of a queue is a movie-booking counter. The person who comes first is serviced first, meaning that he or she gets a chance to book his or her ticket before the next person arrives to the counter.

This is another common data structure used extensively. For example, the producer-consumer design pattern uses a work queue (https://dzone.com/articles/producer-consumer-pattern).

Here is a REPL session to familiarise yourself with queue operations:

```
scala> import scala.collection.immutable.Queue;
import scala.collection.immutable.Queue

scala> val q = Queue[Int]();
q: scala.collection.immutable.Queue[Int] = Queue()

scala> val q1 = q.enqueue(1)
q1: scala.collection.immutable.Queue[Int] = Queue(1)

scala> val q2 = q1.enqueue(2)
q2: scala.collection.immutable.Queue[Int] = Queue(1, 2)

scala> q2.dequeue
res6: (Int, scala.collection.immutable.Queue[Int]) = (1,Queue(2))

scala> val (e, _) = q2.dequeue
e: Int = 1
```

Elements are inserted using the `enqueue` method and removed using the `dequeue` method.

In the next chapter, we will study how the persistent LIFO queues are designed. Adding elements to the queue is *O(1)*. Removing an element is usually *O(1)*, as we will soon see.

Sets

A set is an ADT that does not contain duplicate elements. A very common operation of sets is a membership check.

Here are Scala sets in action:

```
scala> val input = List.range(1, 10) ++ List.range(7, 13)
input: List[Int] = List(1, 2, 3, 4, 5, 6, 7, 8, 9, 7, 8, 9, 10, 11, 12)
scala> val s = Set(input:_*)
```

```
s: scala.collection.immutable.Set[Int] = Set(5, 10, 1, 6, 9, 2, 12, 7, 3,
11, 8, 4)
scala> s(10)
res8: Boolean = true

scala> s(11)
res9: Boolean = true

scala> s(100)
res10: Boolean = false
```

Scala sets are implemented internally as Hash Tries, just like the map ADT. The membership check is a lookup operation on the *trie*, which is very fast. See `https://www.scala-lang.org/docu/files/collections-api/collections_19.html` for more information.

Sorted set

A set does not preserve the order in which elements were inserted. If we need this feature, we need to use a sorted set:

```
scala> val ss = scala.collection.immutable.TreeSet(input:_*)
ss: scala.collection.immutable.TreeSet[Int] = TreeSet(1, 2, 3, 4, 5, 6, 7,
8, 9, 10, 11, 12)

scala> ss(10)
res11: Boolean = true

scala> ss(13)
res12: Boolean = false
```

A sorted set is implemented as a Red-Black tree. A **Red-Black tree** is a balanced Binary Search Tree. By balanced, we mean all the paths from the root of the tree to a leaf have lengths that differ by at most one element.

We look at this intriguing and interesting data structure in a later chapter. Red-Black trees have a guaranteed lookup complexity of $O(log_n)$.

Summary

This was a whirlwind tour of the basics. We started with a look at the *Big O* notation, which is used to reason about how fast an algorithm could run. Next came the notion of *space-time* trade-off.

We saw how trading off some space by caching known results saves time; it avoids needless computations. We looked at pure functions and referential transparency and saw how pure functions are amenable to memoization.

We looked at the idea of *effective constant time* operations.

FP programs use collections heavily. We looked at some common collection operations and their complexities.

We noted how complexity changes in the face of immutability. Building all this ground will give us enough foothold to look at our next data structure, that is, lists.

On to it!

3
Lists

Let's start looking at the first fundamental data structure: lists.

Lists permeate the functional world. LISP is one of the earliest programming languages. The name LISP means *list processor*.

The imperative world also uses lists. In an algorithmic sense, lists are great for growing incrementally, for example, when we append elements to an existing list. List append is an *O(1)* operation in the imperative world. Deleting and inserting nodes anywhere in the list is an *O(1)* operation too. When we insert or delete a node, its predecessor and successor (if any) are the only nodes affected–a few pointers are juggled and the insertion or deletion is done.

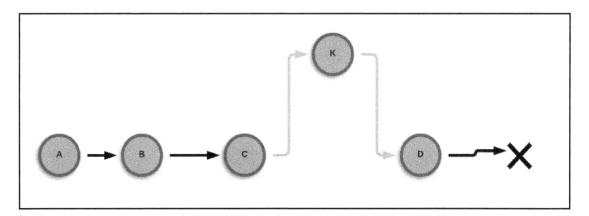

For example, in the preceding list, when we insert node **K**, the algorithm is pretty simple:

```
Set k.next = c.next
Set c.next = k
```

This works! Largely as in the imperative world, mutating a node in place is okay.

In the functional, immutable world, things are pretty different though. In this chapter, we will take a close look at functional list algorithms that are written without updating the list at all. We will also learn about persistent data structures and a few more Scala idioms as we go.

First steps

Let's define the list nodes. First, let's briefly look at a `sealed trait`. A `trait` is just an interface. It may contain methods too. See `http://joelabrahamsson.com/learning-scala-part-seven-traits/` for more information.

The `sealed` keyword allows the compiler to do exhaustive checking. For example, here is an REPL session to see this feature in action:

```
scala> trait A
defined trait A
scala> case class B() extends A
defined class B

scala> case class C() extends A
defined class C
scala> def m(a: A) = a match {
     |    case B() => println("It is a B")
     | }
```

Note the definition of the `m(A)` method. It just handles objects of this type: `B`.

What happens when we call the `m(C())` method? We get a match error:

```
scala> m(C())
scala.MatchError:..
```

The `sealed` keyword helps in this case. Just change the definition of `trait A` as follows:

```
sealed trait A
```

You also need to redefine the B and C case classes. Just repeating the definition should be fine.

Now when you redefine the m method, the compiler generates a warning:

```
scala> sealed trait A
defined trait A

scala> case class B() extends A
defined class B

scala> case class C() extends A
defined class C

scala> def m(a: A) = a match {
     | case B() => println("It is a B")
     | }
<console>:23: warning: match may not be exhaustive.
It would fail on the following inputs: B(), C()
       def m(a: A) = a match {
                        ^
```

The sealed keyword *ensures* that the trait and its implementations are in the same source file. This helps the compiler to perform the *missing case clause* analysis. See http://underscore.io/blog/posts/2015/06/02/everything-about-sealed.html for more information.

Note that the List, as shown in the following code, is not from Scala's standard library; it is our own implementation and is defined as sealed trait:

```
scala>   sealed trait List[+A]
scala>   case object Nil extends List[Nothing]
scala>   case class ::[+A](head: A, tail: List[A]) extends List[A]
```

The following diagram shows a pictorial representation of the node hierarchy:

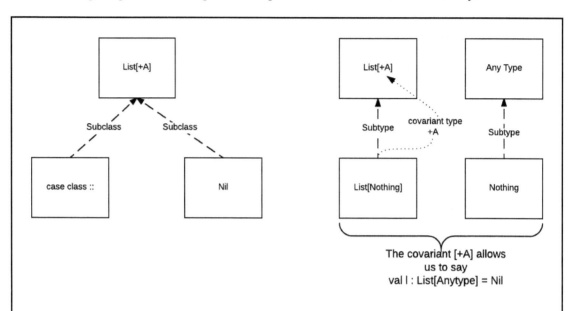

Let's closely look at our `List` definition:

```
sealed trait List[+A]
```

Notice `[+A]`? This makes the `List` type *polymorphic*, that is, it presents the `List` type as `List[Int]`, `List[String]`, and so on.

The + sign in `List[+A]` makes the type a covariant. If `Car` is a subtype of `Vehicle`, then `List[Car]` is a subtype of `List[Vehicle]`. Now refer to the following code:

```
case object Nil extends List[Nothing]
```

Here, `Nil` is a `case` object, the only instance of the `Nil` class. Because of the covariance, we can do the following:

```
scala> val list: List[Int] = Nil
list: ch02.mylist.List[Int] = Nil

scala> val list: List[String] = Nil
list: ch02.mylist.List[String] = Nil
```

Finally, we have a `case` class with a somewhat strange name; the name is `::`. Check out the following code:

```
case class ::[+A](head: A, tail: List[A]) extends List[A]
```

This allows us to define the list as follows:

```
scala> val l1 = ::(1, ::(2, Nil))
l1: ch02.mylist.::[Int] = ::(1,::(2,Nil))
```

Here is a diagram that shows how our list is composed:

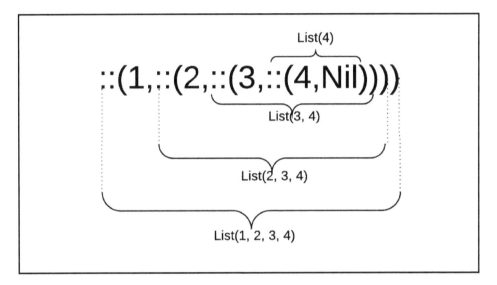

Constructing a list this way is awkward though. Here is the `apply` method that allows us to construct lists easily:

```
def apply[A](as: A*): List[A] =
  if (as.isEmpty) Nil
  else ::(as.head, apply(as.tail: _*))
```

We define the `apply` method in the list's companion object. The method takes a `varargs` argument, indicated by `A*`. It allows us to create a list with a simple syntax: `List(1, 2, 3)`, `List("a", "list", "of", "strings")`, and so on. Now consider the following code:

```
if (as.isEmpty) Nil
```

If there are no arguments–for example, `List()`–then we get a `Nil` list. The list has a `head` element and a list of zero or more elements:

```
else ::(as.head, apply(as.tail: _*))
```

As you can see, we set `head` to the first argument. Scala's `spalt` operator, `_*`, is used to recursively define the `tail`.

This allows us to define lists as follows:

```
scala> val l1 = List(1, 2, 3, 4)
l1: ch02.mylist.List[Int] = ::(1,::(2,::(3,::(4,Nil))))
```

See http://alvinalexander.com/scala/how-to-define-methods-variable-arguments-varargs-fields for more on Scala's `varargs` and `splat` operators.

List head and tail

Here is the simplest `List` method that gets the first element, that is, `head` of the list:

```
scala> def head[A](list: List[A]): A = list match {
     |     case Nil => sys.error("tail of empty list")
     |     case ::(x, _) => x
     |   }
head: [A](list: ch02.mylist.List[A])A
```

We perform a pattern match on the list. The first clause `case Nil => sys.error("tail of empty list")` matches when the list is empty. An empty list will not have a first element so we raise an exception:

```
case ::(x, _) => x
```

This clause matches when there are one or more elements in the list.

The following diagram shows how the pattern match works:

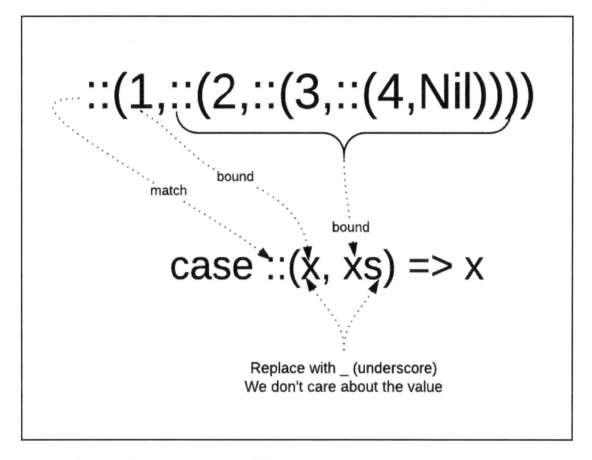

We usually write the second clause as follows:

```
case x :: _ => x
```

This reads better. The expression `::(x, _)` can always be written in infix notation as `x :: _`. See http://danielwestheide.com/blog/2012/11/21/the-neophytes-guide-to-scala-part-1-extractors.html for more information. Here is the revised `head` method:

```
scala> def head[A](list: List[A]): A = list match {
     |    case Nil => sys.error("tail of empty list")
     |    case x :: _ => x
     |  }
```

Getting a list's head is an *O(1)* operation, as we just need to look at the first element (if any) and return it.

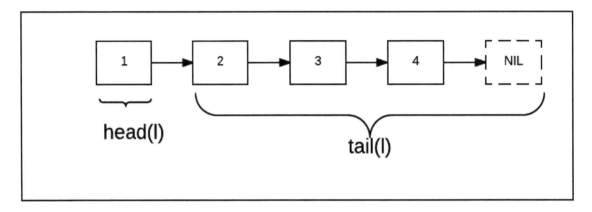

Here is the `tail` method:

```scala
scala>    def tail[A](list: List[A]): List[A] = list match {
     |        case Nil => sys.error("tail of empty list")
     |        case _ :: xs => xs
     |    }
```

The first clause is hit when we invoke `tail` on an empty list. This, just like head, is invalid. Therefore, the clause throws an error.

The second clause case `_ :: xs => xs` returns everything except head. Here is an example of using both these methods:

```scala
scala> val l = List(1, 2, 3, 4, 5)
l: ch02.mylist.List[Int] = ::(1,::(2,::(3,::(4,::(5,Nil)))))
scala> head(l)
res5: Int = 1
scala> val l1 = tail(l)

l1: ch02.mylist.List[Int] = ::(2,::(3,::(4,::(5,Nil))))
```

The complexity of the `tail` method is, again, *O(1)*. We just return the original list's elements, except the first. In other words, we *structurally share* the elements between l (original list) and l1 (tail):

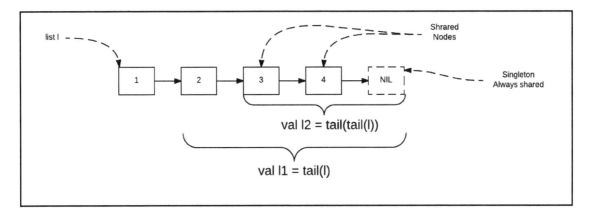

What would happen with `tail(tail(l))`? The elements, except `head`, of the `tail` are again structurally shared.

Drop elements

We are chugging along nicely. Let's look at element removal:

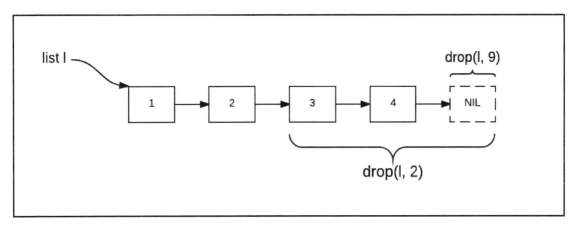

Here comes the method definition:

```
scala> :paste
// Entering paste mode (ctrl-D to finish)

  def drop[A](l: List[A], n: Int): List[A] =
    if (n <= 0) l
    else l match {
```

```
        case Nil => Nil
        case _ :: t => drop(t, n - 1)
    }
```

We drop n (or less than n) elements starting at the beginning of the list. If there are less than n elements, we return a `Nil` (empty) list.

What is the complexity of the `drop` method? It is *O(n)*, as we might end up traversing the entire list.

The `dropWhile` method takes a list and predicate function. It invokes the function on successive elements and drops them if it evaluates to `true`. It stops when the function evaluates to `false`:

```
scala> def dropWhile[A](l: List[A], f: A => Boolean): List[A] = l match {
    |       case x :: xs if f(x) => dropWhile(xs, f)
    |       case _ => l
    |   }
```

Here is an REPL session using it:

```
scala> dropWhile(l, (x: Int) => x <= 2)
res14: ch02.mylist.List[Int] = ::(3,::(4,::(5,Nil)))
```

The following figure shows how structural sharing is central to the `dropWhile` method:

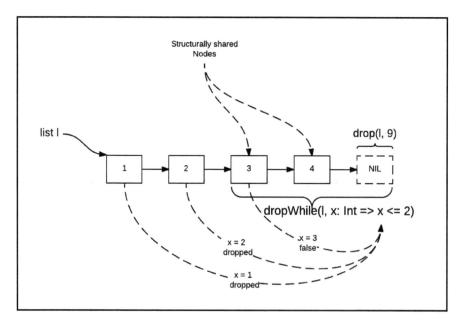

Note that `dropWhile` is a higher order function. It takes a function as an argument. Functions are *first-class values*, just like `Int` or `String`. So we can provide them as arguments to other functions.

Here is a better form, when `dropWhile` is rewritten as a curried function:

```
def dropWhile[A](l: List[A])(f: A => Boolean): List[A] = l match {
  case x :: xs if f(x) => dropWhile(xs)(f)
  case _ => l
}
```

This frees us from writing the x variable's type. It is inferred for us by the compiler:

```
scala> dropWhile(l)(x => x <= 2)
res16: ch02.mylist.List[Int] = ::(3,::(4,::(5,Nil)))

scala> dropWhile(l)(_ <= 2)
res17: ch02.mylist.List[Int] = ::(3,::(4,::(5,Nil)))
```

For more information, see *Programming in Scala, Third Edition* (https://www.safaribooksonline.com/library/view/programming-in-scala/9780981531687/control-abstraction.html).

The complexity of `dropWhile` is $O(n)$.

Concatenating lists

In the imperative world, where we perform mutation as needed, concatenating two linked lists is easy.

Given the two lists **a** and **b**, we just traverse the first list **a** until we reach its last node. Then we change its next pointer to the `head` of list **b**:

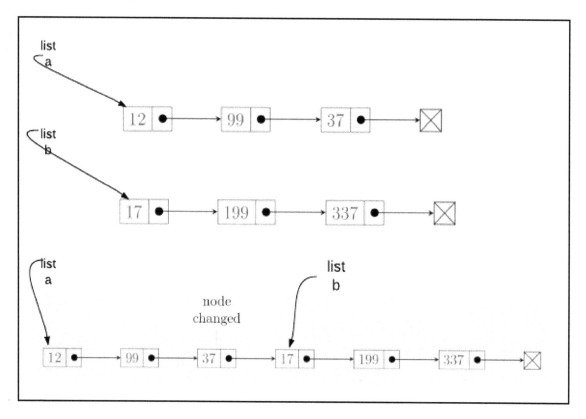

Note what happened to the original list **a**. It changed. The original list simply does not exist anymore.

We destroyed **list a** when we connected its third node to **list b**. The preceding list mutation is also not thread-safe. As seen in the previous chapter, additional mechanism, such as locking, is needed to make sure the state is synchronized correctly.

We could do the concatenation by keeping the original list intact; we do this by copying **list a** into another **list c** and then changing the third node of the new list to point to **list b**.

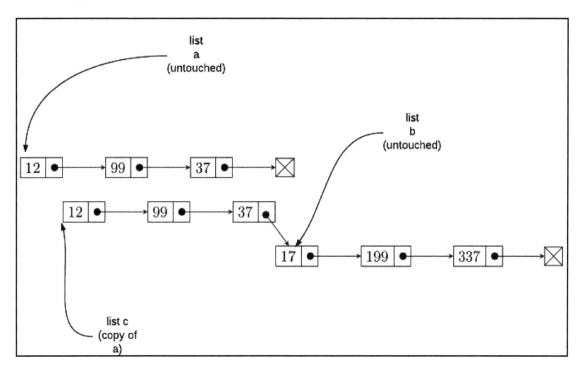

Our **list a** and **list b** are not touched at all. We copy **list a** into a new list, namely **list c**, and change its third node to point to the head of **list b**.

Note that the new list, that is, **list c**, and the existing list, **list b**, share nodes with values **17**, **199**, and **337**.

We trade-off some more memory to maintain multiple versions of the list, thereby avoiding any locking.

Here is the append method:

```scala
scala>   def append[A](a1: List[A], a2: List[A]): List[A] =
     |       a1 match {
     |         case Nil => a2
     |         case h :: t => ::(h, append(t, a2))
     |       }
```

The first case case Nil => a2 matches when we reach the end of the first list, that is, a1.

Here's an example of how to use this method:

```scala
scala> append(List(1,2), List(3,4))
res18: ch02.mylist.List[Int] = ::(1,::(2,::(3,::(4,Nil))))
```

The complexity of the `append` method is proportional to the length of the list being copied.

Persistent data structures

If we follow the second *copy-as-much-needed-and-share* strategy, a thread holding a reference to the original lists will never be surprised by any changes. From the thread's point of view, nothing has changed and things continue as before.

However, the change needs to happen somewhere. How else would we grow/shrink data structures? The change does indeed happen, but only at creation time.

Let's look at what we mean when we say *creation time*. Consider the following snippet:

```scala
scala> :paste

@scala.annotation.tailrec
def print(low: Int, high: Int): Unit =
  if (low > high)
    println("Done")
  else {
    println(low)
    print(low+1, high)
  }
scala> print(1, 10)
1
2
...
10
Done
```

We are printing a range of numbers. As we know, this needs some kind of looping. However, there is no counter variable that is getting changed (in other words, mutated).

The change happens just before we start the next recursive call. The loop is still there; however, it is expressed using recursion. Also note the use of the `@scala.annotation.tailrec` annotation to make sure the tail call optimization kicks in and the stack does not overflow.

Algorithms that follow the *copy-and-share* strategy lead us to persistent data structures. Such data structures always *preserve* their earlier form. Any update operations on these data structures do not affect the structure already in place.

The term "persistence" is overloaded in practice. The persistence we just covered in no way refers to saving anything on a disk. An application's persistence layer is a very different thing and we do not refer to it.

We will see a lot of persistent data structures in this book, all depicting the property of preserving their existing form upon an update operation.

This is essentially the copy-on-write way of sharing state. We keep sharing until someone attempts to change the shared state. Then, the shared state is copied and the change is made without disturbing the original state.

See `http://stackoverflow.com/questions/628938/what-is-copy-on-write` for a very good discussion on this technique.

Tail call optimization

We are using recursion to express looping. However, recursive code could result in a stack overflow. We need to understand an important optimization technique called **tail call optimization** (**TCO**). When TCO is applied, behind the scenes, recursion is expressed as a loop. This avoids stack frames, and subsequently, stack overflow.

For example, here is the preceding `print` method modified to print the range in reverse:

```
scala> def revprint(low: Int, high: Int): Unit = {
     |    if (low <= high) {
     |       revprint(low+1, high)
     |       println(low)
     |    }
     | }
scala> revprint(1,4)
4
3
2
1
```

The following diagram shows the stack frames used:

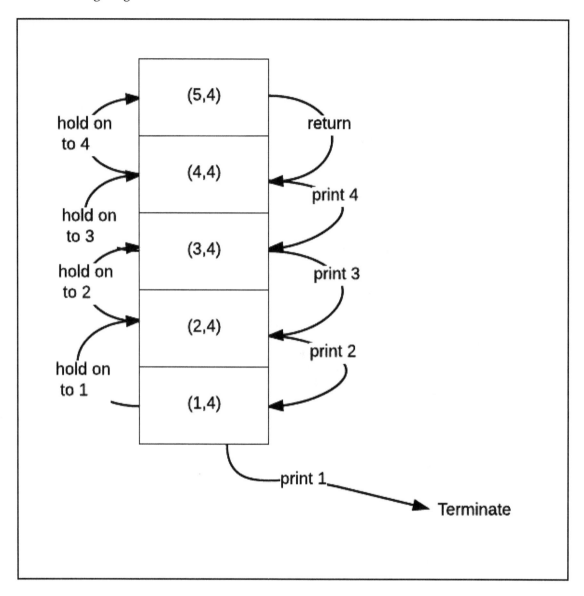

We need the stack to remember the value of `low` as we need to print it. We remember this value on a *stack frame*. There are limited stack frames though. So if we call `revprint` as `revprint(1, 10000)`, then we get `StackOverflowError`.

If we can express our algorithm such that the recursive call would be in the tail position, Scala would rewrite the recursion as a loop. A call in the tail position does nothing other than return the value.

We still express our loops using recursion and immutability and still don't overflow the stack. You can have your cake and eat it too!

The algorithm could be changed so the recursive call is in the tail position, as shown:

```scala
scala> def revprint_tco(low: Int, high: Int): Unit = {
     |    if (high >= low) {
     |      println(high)
     |      revprint_tco(low, high-1)
     |    }
     | }
```

When we say a recursive call is in the *tail position*, it means it does nothing other than return the result, if any. Note that the algorithm doesn't need to hold on to any state. Scala applies the TCO optimization here by default.

However, to make sure TCO kicks in and all the calls are indeed in the tail position, we use the @tailrec annotation. If we try to apply the annotation to the revprint method, we get an error. The annotation works fine for the revprint_tco method.

List append

Consider appending a node to a list. In the mutation world, we traverse the list until we reach the end and then change the last node to point to the new node. This is costly when the list is long and has a complexity of *O(n)*.

For a persistent list (immutable and structurally shared) appending a new value, we need to traverse until the end of the list, copying all the elements on the way.

However, as noted, appending to a list is anyway a slow operation. When we need to append values, we need to ask ourselves whether lists are the right fit for the problem.

Whenever we want to grow a list by appending to the end, we should instead use a vector. When we are done with all of the appending, we could convert the vector into a list, if needed.

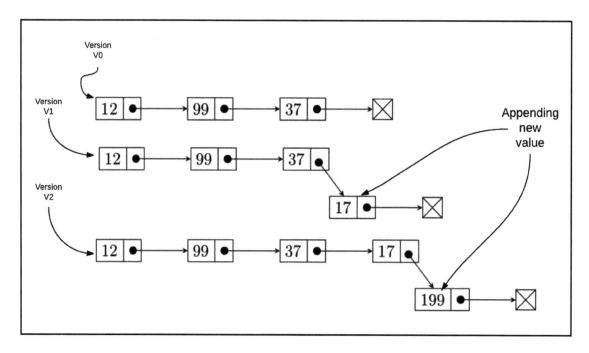

We can look at the original list at **V0**. This list has three nodes, holding the values **12**, **99**, and **37**.

When we append the value **17**, the original three nodes are copied, and then *at construction time*, the node with the value **17** is added. The data structure is at **V1**.

Next, when we append the value **199** to this list, all the four existing nodes are copied and the value **199** is added, thereby yielding a new list. Now the data structure is at **V2**.

All the data structure versions coexist via structural sharing. The data structure is thus *persistent*.

Here is the `appendElem` method:

```
scala>   def appendElem[A](l: List[A], elem: A): List[A] = l match {
     |        case Nil => List(elem)
     |        case x :: xs => ::(x, appendElem(xs, elem))
     |    }
scala> l
res5: ch02.mylist.List[Int] = ::(1,::(2,::(3,::(4,Nil))))

scala> appendElem(l, 5)
res6: ch02.mylist.List[Int] = ::(1,::(2,::(3,::(4,::(5,Nil)))))
```

The first clause matches when the list is `Nil`:

```
case Nil => List(elem)
```

We just return a list of one element: `elem`. The second clause is as follows:

```
case x :: xs => ::(x, appendElem(xs, elem))
```

It matches when the list has at least one element. We hold on to x and recursively call the `appendElem` method again with the `tail` of the list.

Eventually, we hit `Nil`, get back `List(elem)`, and stick it in place of `Nil`, thereby growing the list by one element.

As we need to copy each element to preserve the earlier version, `appendElem` has a runtime complexity of *O(n)*; space complexity is *O(n)* too.

List prepend

Note that lists are great when we insert a node at the beginning, in other words, prepend a value to the `head` of a list. Let's see how that works.

We have the list with values **17**, **199**, and **337**. We prepend the value **37** first. Next, we prepend **99** to the resulting list. Finally, we prepend **12** to the new resulting list again.

When we prepend the value **37**, we just allocate the node; while constructing this node, we append the original list to **37**. In other words, **37** becomes the head and the original list becomes the tail.

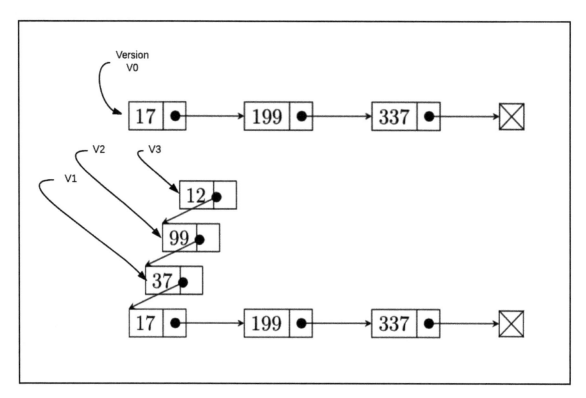

Note that there is no copying needed at all. We just allocate the new node and append the existing list to it. As this does not affect the persistence of the already existing data structure, we have a very efficient prepend operation.

The version **V0** is, as before, **list b**. **V1** gets created when **37** is prepended to the list, **V2** when **99** is prepended, and **V3** when **12** is prepended. Thus, the complexity of prepend is *O(1)*.

Here is the prepend method:

```
scala>    def prepend[A](list: List[A], v: A) = list match {
      |         case Nil => ::(v, Nil)
      |         case x :: xs => ::(v, list)
      |    }
scala> prepend(l, 0)
res9: ch02.mylist.::[Int] = ::(0,::(1,::(2,::(3,::(4,Nil)))))
```

The following figure shows how structural sharing is at the heart of the prepend method:

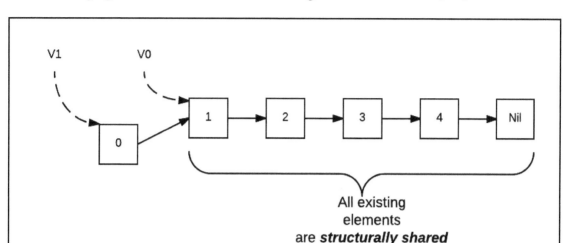

The first clause is hit when the list is Nil. We just create List(elem) and return it. The second clause case x :: xs => ::(v, list) is hit when the list has at least one element (it could have been more).

We just create a new list with v as head and the argument list as tail.

Getting value at index

We cannot do random indexing on a linked list. We have to traverse the list and keep counting until we reach the element or are presented with an error.

Before we look at the method, let's look at Scala's tuple matching. For example, consider the following snippet:

```
scala> val tup = (1, 1)
tup: (Int, ch02.mylist.List[Int]) = (1,::(1,::(2,::(3,::(4,Nil)))))

scala> tup match {
     |    case (i, x :: xs) => s"i = ${i}, x = ${x}"
     | }
<console>:19: warning: match may not be exhaustive.
It would fail on the following input: (_, Nil)
          tup match {
```

```
         ^
    res20: String = i = 1, x = 1
```

The warning is issued, as we are not handling the case, when the second element of the tuple is an empty list. As the preceding code snippet is merely illustrative, we can live with it.

The following diagram shows how tuple matching works:

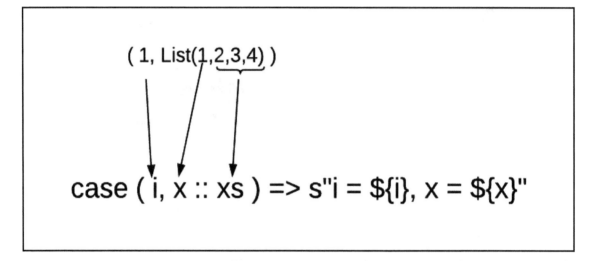

We have a tuple `tup` holding `Int` and `List[Int]`. As shown in the diagram, the match binds the first element to the variable `i`. The variable `x` is bound to the nested list's first element.

Here is a quiz for you:

- Change the code so `x` is bound to the *second element* of the list.

Once we understand nested matching, understanding the `elementAtIndex` method will become simpler:

```
scala>    def elemAtIndex[A](l: List[A], i: Int): A = (1,i) match {
     |        case (Nil, _) => sys.error(s"index ${i} not valid")
     |        case (x :: xs, 0) => x
     |        case (x :: xs, _) => elemAtIndex(xs, i-1)
     |    }
scala> val l = List(1,2,3,4)
l: ch02.mylist.List[Int] = ::(1,::(2,::(3,::(4,Nil))))
```

```
scala> elemAtIndex(l,1)
res15: Int = 2

scala> elemAtIndex(l,3)
res16: Int = 4
```

We match the tuple (List[A], Int). The first clause, case (Nil, _) =>
sys.error(s"index ${i} not valid") is hit when the list is Nil. We don't even care
about the index, so just use _ as a placeholder.

The second clause case (x :: xs, 0) => x is hit when the list is not empty and the
index is 0. We return x as the value desired.

The third clause, case (x :: xs, _) => elemAtIndex(xs, i-1) is hit when the list is
not empty and the index is non-zero. We call the method again with tail and the
decremented index value.

The complexity of the method is *O(n)*. Here are a couple of questions for you (quiz time):

- Is the elemAtIndex method tail recursive?
- What if we call the method with a negative index? If there is any issue, fix the
  code to handle it.

Modifying a list value

What happens when we try to modify a list value? As before, copying and sharing kicks in
to make the data structure persistent.

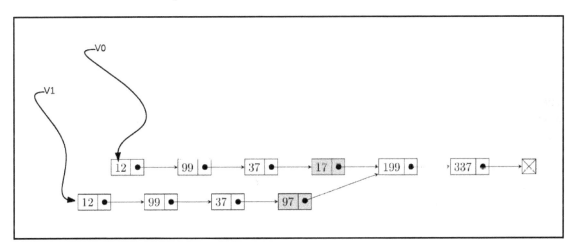

We changed the fourth node's value from **17** to **97**. To keep the persistent view correct, for **V1**, we had to copy all the nodes on the path leading up to the modified node.

On the other hand, nodes on all other paths are shared. The case with node deletion is also the same.

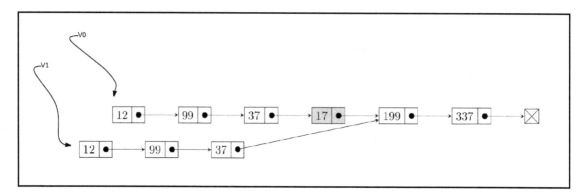

Note that the copying process is the same as the earlier node modification case.

The insertion case would be similar as well. Play with the scenario where we add node **52** after the second node with value **99**:

```
scala>    def setElem[A](l: List[A], i: Int, elem: A): List[A] = (l, i)
match {
    |      case (Nil, _) => sys.error(s"index ${i} not valid")
    |      case (_ :: xs, 0) => ::(elem, xs)
    |      case (x :: xs, _) => ::(x, setElem(xs, i-1, elem))
    |    }
scala> val l = List(12, 99, 37, 17, 199, 337)
l: ch02.mylist.List[Int] = ::(12,::(99,::(37,::(17,::(199,::(337,Nil))))))

scala> setElem(l, 3, 97)
res13: ch02.mylist.List[Int] =
::(12,::(99,::(37,::(97,::(199,::(337,Nil))))))
```

The first clause, `case (Nil, _) => sys.error(s"index ${i} not valid")` is hit when the index is invalid. We have seen this case before–we are trying to set a non-existent index element. This results in an error.

The second clause, `case (_ :: xs, 0) => ::(elem, xs)` does the actual *non-destructive* update. We create a new node in place of the existing one and return it, thereby terminating the traversal.

The final clause, `case (x :: xs, _) => ::(x, setElem(xs, i-1, elem))` is hit when we reach an earlier node and need to continue down the list. We copy the current node and continue the traversal by invoking `setElem` recursively.

Here is a quiz for you:

- The `setElem` method is not tail-recursive, and as a result, it will crash for very long lists. How would you change the code to handle this case?

Summary

We looked at lists, the basic data structures used for functional programming. We had a detailed look at how list algorithms work in the immutable, side-effect-free functional world.

We saw the notion of a persistent data structure wherein the original version of the data structure is never mutated. Instead, we created a new structure, reflecting the change. We saw many cases of node insertion and removal for both lists and binary trees.

All of this copying could be thought of as too expensive. However, as we saw, we shared as many nodes as possible with the original data structure. We need to copy nodes only when we need to preserve the original version.

We implemented lists in Scala with the view of studying persistence and sharing. We implemented some functional list algorithms to better understand the fundamental concepts at play. In the rest of the book, we will use Scala's immutable lists.

Hope you have enjoyed the journey so far. Let's continue the fun ride and look at binary trees!

4
Binary Trees

A **tree** is a data structure that simulates a hierarchical tree structure with a *root* node and *children* nodes. A child node can have children nodes of its own, thus realizing the hierarchy. Just like a list, a tree is defined *recursively*.

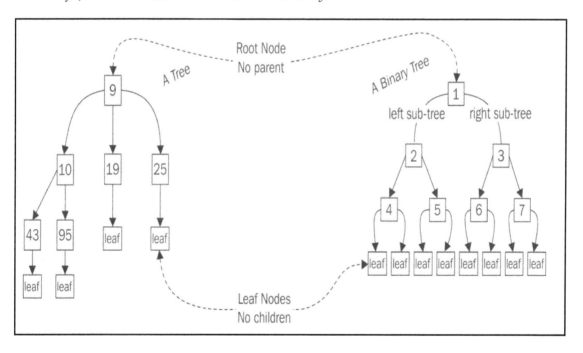

A *binary tree* is a tree in which each node has zero, one, or at most two child nodes. See http://opendatastructures.org/ods-java/6_Binary_Trees.html for an excellent refresher on binary trees.

In this chapter, we will look at the functional implementation of binary trees. As in the previous chapter, we will not perform any *in-place* mutations. Any update operation will preserve the earlier version, thereby making it a *persistent* binary tree.

We will implement many common binary tree algorithms. After covering binary trees, we will look at Binary Search Trees.

Node definitions

Similar to lists, our binary tree is a trait, BinTree[+A]:

```
sealed trait BinTree[+A]
case object Leaf extends BinTree[Nothing]
case class Branch[A](value: A, left: BinTree[A], right: BinTree[A]) extends
BinTree[A]
```

The sealed trait BinTree[+A] array defines a *sealed* trait. As it is sealed, we can extend it only in the same source file. We saw in Chapter 3, Lists how this helps the compiler to check for *exhaustive pattern matching*:

```
case object Leaf extends BinTree[Nothing]
```

The Leaf node is a terminator node, just like we have the Nil node in lists. Just like Nil, Leaf is a case object, as we just need only one instance of it:

```
case class Branch[A](value: A, left: BinTree[A], right: BinTree[A]) extends
BinTree[A]
```

The Branch node holds a value, of type A, and a left and right subtree. These subtrees could be either branches or leaves.

Thus, we define the binary tree in terms of itself; in other words, it is a recursively defined structure, similar to `List`:

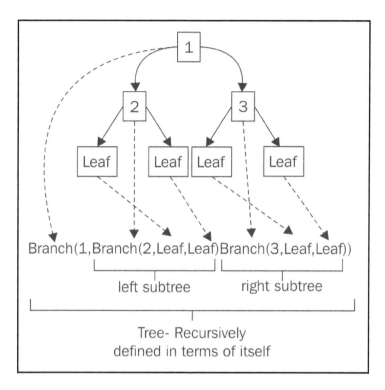

 Note that this is just a binary tree. We will look at its specialization, BST, pretty soon.

Building the tree

Let's try building one such tree. The input values are fed from a `List`. Note that we are using Scala's immutable lists:

```scala
scala>    def buildTree[A](list: List[A]): BinTree[A] = list match {
     |        case Nil => Leaf
     |        case x :: xs => {
     |          val k = xs.length / 2
     |          Branch(x, buildTree(xs.take(k)), buildTree(xs.drop(k)))
     |        }
     |    }
buildTree: [A](list: List[A])BinTree[A]

val treeFromList = buildTree(list)
treeFromList: BinTree[Int] =
Branch(1,Branch(2,Branch(3,Leaf,Leaf),Branch(4,Leaf,Leaf)),Branch(5,Branch(
6,Leaf,Leaf),Branch(7,Leaf,Branch(8,Leaf,Leaf))))
```

The method creates a balanced binary tree. Such a balanced tree's left and right subtrees have almost equal number of nodes.

The first case is simple:

```scala
case Nil => Leaf
```

The clause is hit when the list is empty. We just return a `Leaf` node in this case.

The second clause is as follows:

```scala
case x :: xs => {
  val k = xs.length / 2
  Branch(x, buildTree(xs.take(k)), buildTree(xs.drop(k)))
}
```

This clause is hit when the list has *one or more* elements. We take the length of the *tail list* and divide it by 2. The head and the first half of the tail list form the left subtree. The tail list elements on the right-hand side form the right subtree.

As we keep distributing the elements evenly, the tree remains balanced.

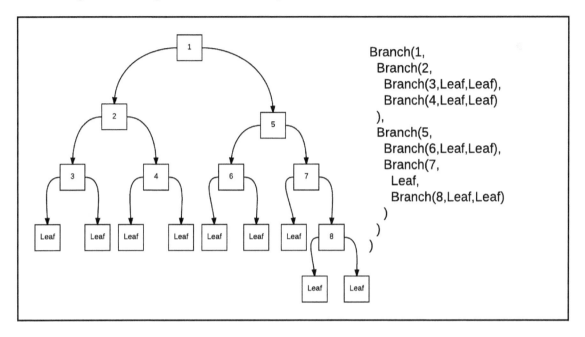

Note that the *pointers are set up at construction time only*. We do *not mutate* the existing data structure at any point.

Here is an image tracing the tree being built:

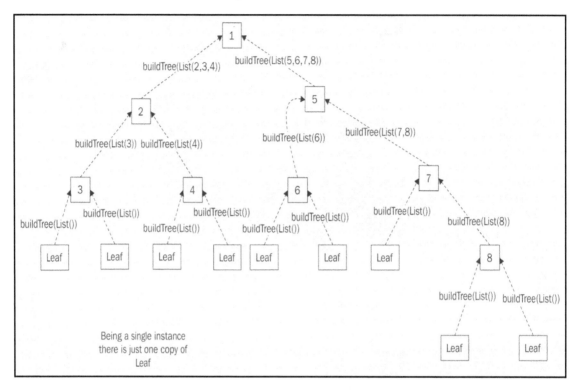

This diagram shows the recursive calls and the list argument each call receives. Please trace out the preceding code with some other list of your choice, as this concept is really fundamental to understanding the rest of the book.

Size and depth

The `size` method computes the number of *non-leaf* nodes in a binary tree:

```scala
scala>    def size[A](tree: BinTree[A]): Int = tree match {
     |        case Leaf => 0
     |        case Branch(_, l, r) => 1 + size(l) + size(r)
     |    }
...
scala> val list = List(1,2,3,4,5,6,7,8)
...
scala>    val treeFromList = size(buildTree(list))
res7: Int = 8
```

The first clause is hit when the tree is a `Leaf` node:

```
case Leaf => 0
```

We return 0 as the size of `Leaf` is 0 by definition.

The second clause is as follows:

```
case Branch(_, l, r) => 1 + size(l) + size(r)
```

This clause is hit when the we match a `Branch` node. We don't care about `value` as it does not participate in the algorithm.

As mentioned in previous chapter, we ignore it by specifying an underscore (_) character for the `value` part.

The depth of the tree is the length of the longest path from a root to a leaf.

Here is the `depth` method:

```
scala>    def depth [A](tree: BinTree[A]): Int = tree match {
     |        case Leaf => 0
     |        case Branch(_, l, r) => 1 + (depth(l) max depth(r))
     |    }
depth: [A](tree: BinTree[A])Int
```

This is how you exercise both these methods:

```
scala> val list = List(1,2,3,4,5,6,7,8)
...
scala>    val treeFromList = buildTree(list)
...
scala> size(treeFromList)
res7: Int = 8
scala> depth(treeFromList)
res8: Int = 4
```

The `size` method returns 8, whereas the `depth` method returns 4.

Complete binary trees

If a binary tree has $size(tree) == 2^{(depth(tree))}-1$, then it is a complete binary tree.

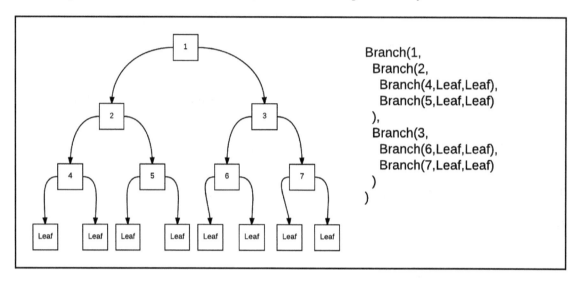

The figure shows a binary tree. Its depth is **3** and size is **7**. As the preceding formula holds ($7 == 2^3-1$), the tree is indeed complete.

Here is a way to generate such a tree:

```
scala> :paste
// Entering paste mode (ctrl-D to finish)

   def buildCompleteTree(v: Int, depth: Int): BinTree[Int] =
      if (depth == 0) Leaf
      else Branch(v, buildCompleteTree(2*v,  depth-1),
buildCompleteTree(2*v+1,  depth-1))
scala> val completeTree = buildCompleteTree(1, 3)
completeTree: BinTree.BinTree[Int] =
Branch(1,Branch(2,Branch(4,Leaf,Leaf),Branch(5,Leaf,Leaf)),Branch(3,Branch(
6,Leaf,Leaf),Branch(7,Leaf,Leaf)))
```

The first check is to see if the depth is 0:

```
if (depth == 0) Leaf
```

If so, we return a `Leaf` otherwise, the `else` clause is hit:

```
    else Branch(v, buildCompleteTree(2*v,  depth-1),
  buildCompleteTree(2*v+1,  depth-1))
```

 Note that we never explicitly initialize a subtree with a `Leaf` node. Also, for both the subtrees, *depth is decreased by one*. So except the values, we generate the same `Branch` nodes for the left and right subtrees.

For more information on complete binary trees, please see:

http://stackoverflow.com/questions/12359660/difference-between-complete-binary-tree-strict-binary-tree-full-binary-tre

Comparing trees

How do we compare two trees to know whether they are equal or not? Note that the trees could have the same values but could differ structurally.

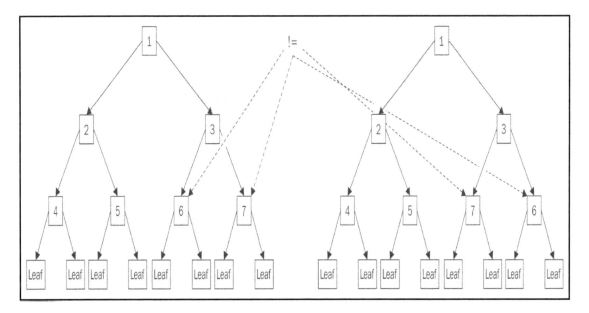

The diagram shows two complete binary trees. However, as shown, these two are not equal.

To check this, we need to *traverse both the trees at the same time*. How could we perform this traversal? Of course, we do this using a *tuple match*:

```
scala>   def equal[A](tree1: BinTree[A], tree2: BinTree[A]): Boolean =
(tree1,
     |        tree2) match {
     |        case (Leaf, Leaf) => true
     |        case (Branch(v1, l1, r1), Branch(v2, l2, r2)) if v1 == v2 =>
     |          equal(l1, l2) && equal(r1, r2)
     |        case _ => false
     |    }
scala>   val tree1 = buildTree(List(1,2,3,4,5,6,7))
tree1: BinTree.BinTree[Int] =
Branch(1,Branch(2,Branch(3,Leaf,Leaf),Branch(4,Leaf,Leaf)),Branch(5,Branch(
6,Leaf,Leaf),Branch(7,Leaf,Leaf)))

scala>   val tree2 = buildTree(List(1,2,3,4,5,7,6))
tree2: BinTree.BinTree[Int] =
Branch(1,Branch(2,Branch(3,Leaf,Leaf),Branch(4,Leaf,Leaf)),Branch(5,Branch(
7,Leaf,Leaf),Branch(6,Leaf,Leaf)))

scala> equal(tree1, tree2)
res0: Boolean = false

scala> equal(tree1, tree1)
res1: Boolean = true
```

 Note that we just check the *absolute minimum* conditions. For example, while traversing, we should hit the *same types* of nodes *at the same time*.

The first clause is as follows:

```
case (Leaf, Leaf) => true
```

This is hit when both the trees are `Leaf`. A `Leaf` is always equal to another `Leaf` as there is only one instance of `Leaf`. It is a `case` object after all!

The second clause is as follows:

```
case (Branch(v1, l1, r1), Branch(v2, l2, r2)) if v1 == v2 =>
     |          equal(l1, l2) && equal(r1, r2)
```

This is hit when we hit two branches at the same time. In this case, we just have a *pattern guard* that checks whether both the `Branch` values are equal. If so, we recursively check both the left and right subtrees.

The third clause is as follows:

```
case _ => false
```

This is hit when none of these two conditions are true. This is because the *patterns are matched in the order they are specified*.

The third clause will be hit, for example, when one tree node is `Leaf` and the other is `Branch`. Another case is when both are `Branch` nodes but the values differ.

We just express the correctness condition first and then a *catchall*; if this is hit, it simply evaluates to `false`.

Flipping a binary tree

This operation exchanges the left and right subtrees *recursively*. Here is a diagram showing a binary tree being flipped:

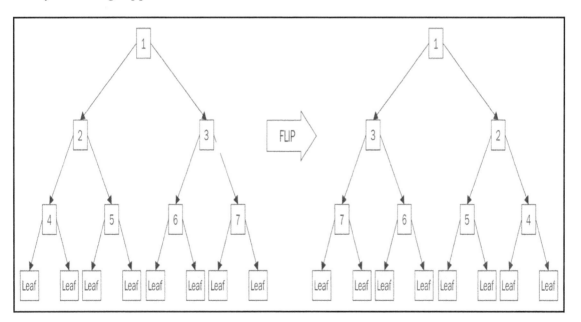

Here is the `flip` method:

```
scala>   def flip[A](tree: BinTree[A]): BinTree[A] = tree match {
    |        case Leaf => Leaf
    |        case Branch(v, l, r) => Branch(v, flip(r), flip(l))
    |    }

scala> val t = buildTree(List(1,2,3,4,5,6,7))
t: BinTree.BinTree[Int] =
Branch(1,Branch(2,Branch(3,Leaf,Leaf),Branch(4,Leaf,Leaf)),Branch(5,Branch(
6,Leaf,Leaf),Branch(7,Leaf,Leaf)))

scala> flip(t)
res5: BinTree.BinTree[Int] =
Branch(1,Branch(5,Branch(7,Leaf,Leaf),Branch(6,Leaf,Leaf)),Branch(2,Branch(
4,Leaf,Leaf),Branch(3,Leaf,Leaf)))
```

It is so simple; we just have two cases. The first case is as follows:

```
case Leaf => Leaf
```

This matches when the tree is a `Leaf`. A `Leaf` node does not have any subtrees, so we just return it. There is nothing to be done here.

The second clause is as follows:

```
case Branch(v, l, r) => Branch(v, flip(r), flip(l))
```

This matches when we hit a `Branch` node. We just *keep* the value as is; however, we *exchange* the *flipped* left and right subtrees.

Here is an interesting method, namely `flippedEqual`, that checks whether the second tree is the flipped form of the first:

```
scala>    def flippedEqual[A](tree1: BinTree[A], tree2: BinTree[A]): Boolean
= (tree1,
    |        tree2) match {
    |        case (Leaf, Leaf) => true
    |        case (Branch(v1, l1, r1), Branch(v2, l2, r2)) if v1 == v2 =>
    |          flippedEqual(l1, r2) && flippedEqual(l2, r1)
    |        case _ => false
    |    }
flippedEqual: [A](tree1: BinTree.BinTree[A], tree2:
BinTree.BinTree[A])Boolean

scala> val t1 = buildTree(List(1,2,3,4,5,6,7))
t1: BinTree.BinTree[Int] =
Branch(1,Branch(2,Branch(3,Leaf,Leaf),Branch(4,Leaf,Leaf)),Branch(5,Branch(
```

```
6,Leaf,Leaf),Branch(7,Leaf,Leaf)))

scala> val t2 = flip(t1)
t2: BinTree.BinTree[Int] =
Branch(1,Branch(5,Branch(7,Leaf,Leaf),Branch(6,Leaf,Leaf)),Branch(2,Branch(
4,Leaf,Leaf),Branch(3,Leaf,Leaf)))

scala> println(flippedEqual(t1, t2))
true

scala> println(flippedEqual(t1, t1))
false
```

I leave the process of tracing out the execution as an exercise.

Binary tree traversal

Binary trees have three well-known traversal methods. We traverse a data structure and *visit* each node to perform some action. For example, we need to visit each `Branch` node to print the value contained within it.

Here is a tree that illustrates all the three types of traversals:

```
scala> val t = Branch(1,Branch(2,Leaf,Leaf),Branch(5, Branch(9,Leaf,Leaf),
Leaf))
...
```

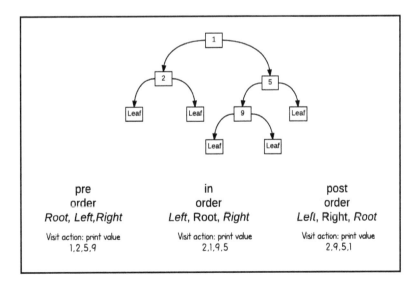

A `preorder` traversal first visits the node *itself,* then its *left* subtree, and eventually it's *right* subtree:

```
scala>    def preorder[A](tree: BinTree[A]): List[A] = tree match {
     |        case Leaf => Nil
     |        case Branch(v, l, r) => v :: (preorder(l) ++ preorder(r))
     |    }
preorder: [A](tree: BinTree.BinTree[A])List[A]
scala> println(preorder(t))
List(1, 2, 5, 9)
```

The first clause is as follows:

```
case Leaf => Nil
```

This is hit when we hit a `Leaf`. Visiting a `Leaf` node does not mean anything, so we just return an empty list: `Nil`.

The second clause is as follows:

```
case Branch(v, l, r) => v :: (preorder(l) ++ preorder(r))
```

This is hit when we hit a `Branch` node. We first take the value and then *prepend it to the result* of recursively calling `preorder` on the left and right subtrees. Note that we are using ++ to concatenate two lists. Here is a snippet of REPL to understand what is going on:

```
scala> List(1,2,3) ++ List(4,5,6)
res16: List[Int] = List(1, 2, 3, 4, 5, 6)

scala> 0 :: (List(1,2,3) ++ List(4,5,6))
res17: List[Int] = List(0, 1, 2, 3, 4, 5, 6)
```

An *in-order traversal* visits the *left* subtree first, then visits the node *itself,* and finally visits the *right* subtree:

```
scala>    def inorder[A](tree: BinTree[A]): List[A] = tree match {
     |        case Leaf => Nil
     |        case Branch(v, l, r) => inorder(l) ++ ( v :: inorder(r))
     |    }
inorder: [A](tree: BinTree.BinTree[A])List[A]

scala> println(inorder(t))
List(2, 1, 9, 5)
```

The first clause analysis is the same as in the `preorder` case. The second clause is as follows:

```
case Branch(v, l, r) => inorder(l) ++ ( v :: inorder(r))
```

This is interesting. We *first* take `inorder(l)`, then take the node value `v`, and finally `inorder(r)`.

A post-order traversal visits the left subtree, then visits the right subtree, and finally the node itself:

```
scala>   def postorder[A](tree: BinTree[A]): List[A] = tree match {
     |       case Leaf => Nil
     |       case Branch(v, l, r) => postorder(l) ++ postorder(r) ++ List(v)
     |   }
postorder: [A](tree: BinTree.BinTree[A])List[A]

scala> println(postorder(t))
List(2, 9, 5, 1)
```

The way `postorder` works is similar to the earlier traversal methods. In the second clause, we traverse the left and right subtrees and then visit the root node.

The accumulator idiom

These methods are easy to understand. However, for *unbalanced trees, concatenating long lists could be slow, very slow.*

We could eliminate list concatenation by sending the result list as an *accumulator* argument:

```
scala>   def preorderAcc[A](tree: BinTree[A], acc: List[A]): List[A] = tree
match {
     |       case Leaf => acc
     |       case Branch(v, l, r) => v :: preorderAcc(l, preorderAcc(r, acc))
     |   }
scala> println(preorderAcc(t, Nil))
List(1, 2, 5, 9)
```

The method now takes an additional argument: `acc`. We *always* set it to `Nil` when we call the method.

As in `preorder`, we have two cases.

The first clause is as follows:

```
case Leaf => acc
```

This just returns the already accumulated values, if any.

The second clause is as follows:

```
case Branch(v, l, r) => v :: preorderAcc(l, preorderAcc(r, acc))
```

This prepends the value v to the *result* of calling `preorder` on both the subtrees:

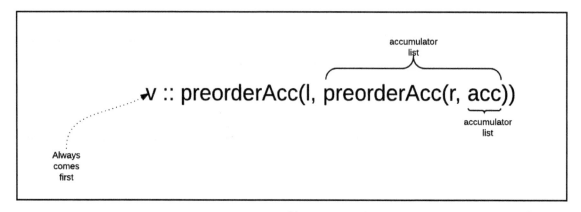

A nice way to trace out how the accumulator grows is to add a judicious `print` statement.

Here is one such trace:

```
scala>   def preorderAcc[A](tree: BinTree[A], acc: List[A]): List[A] = tree
match {
    |           case Leaf => acc
    |           case Branch(v, l, r) => {
    |             println(s"When at ${v} - acc = ${acc}")
    |             v :: preorderAcc(l, preorderAcc(r, acc))
    |           }
    |        }
preorderAcc: [A](tree: BinTree.BinTree[A], acc: List[A])List[A]

scala>   val t1 = buildCompleteTree(1, 3)
t1: BinTree.BinTree[Int] =
Branch(1,Branch(2,Branch(4,Leaf,Leaf),Branch(5,Leaf,Leaf)),Branch(3,Branch(
6,Leaf,Leaf),Branch(7,Leaf,Leaf)))

scala>   println(preorderAcc(t1, List()))
When at 1 - acc = List()
When at 3 - acc = List()
```

```
When at 7 - acc = List()
When at 6 - acc = List(7)
When at 2 - acc = List(3, 6, 7)
When at 5 - acc = List(3, 6, 7)
When at 4 - acc = List(5, 3, 6, 7)
List(1, 2, 4, 5, 3, 6, 7)
```

Here is the trace depicted diagrammatically:

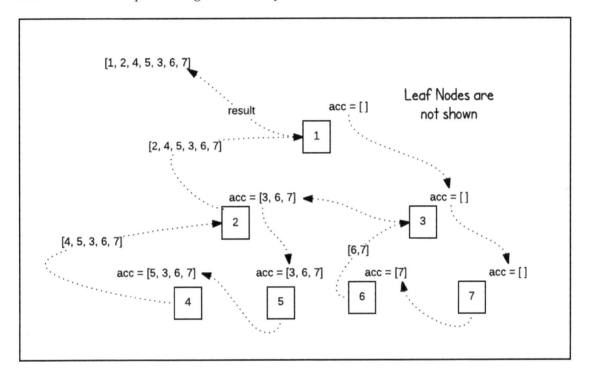

The accumulator versions of `inorder` and `postorder` are shown as follows:

```
scala>  def inorderAcc[A](tree: BinTree[A], acc: List[A]): List[A] = tree
match {
     |     case Leaf => acc
     |     case Branch(v, l, r) => inorderAcc(l, v :: inorderAcc(r, acc))
     |   }
inorderAcc: [A](tree: BinTree.BinTree[A], acc: List[A])List[A]

scala>  def postorderAcc[A](tree: BinTree[A], acc: List[A]): List[A] =
tree match {
     |     case Leaf => acc
     |     case Branch(v, l, r) => postorderAcc(l, postorderAcc(r, v ::
acc))
```

```
    |    }
postorderAcc: [A](tree: BinTree.BinTree[A], acc: List[A])List[A]
```

I leave the process of drawing an accumulator trace of the preceding methods as an exercise.

Binary Search Trees

A **Binary Search Tree** (**BST**) is a binary tree with the following additional property. The value at the root node is *greater (or equal) than all the values in the left subtree*. Likewise, the value is *lesser than (or equal) all the values in the right subtree*.

We keep things simple and don't consider the multiple equal values case. Rather, we implement *dictionaries* using binary trees.

A dictionary is a *list of* (*key, value*) pairs. A key can occur only once, that is, the key is *unique*. For example, we could use dictionaries to compute the count of words in a given text input.

The word count algorithm is simple:

```
words[] = split a string on space characters.
for each word, search the dictionary if it is already present.
if not in dictionary, insert (word, 1)
if found in dictionary, take associated count, cnt
update (word, cnt+1)
```

The following figure shows an abstract dictionary on the left-hand side. The right-hand side of the diagram shows how we could realize it using a BST:

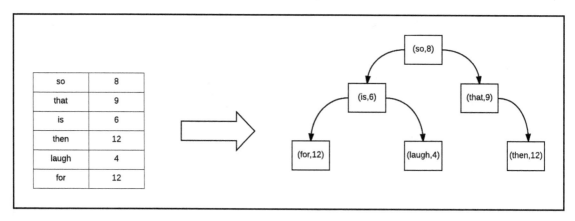

If we just store the words and their counts as a *list of pairs*, for long word lists, it will be slow. As we know, lists give linear access, so *searching* for a word could have $O(n)$ complexity. For all the words in the dictionary, the algorithm complexity would be $O(n^2)$.

Instead, by implementing the dictionary as a BST, let's perform a binary search:

```
type Dictionary[A] = BinTree[(String, A)]
```

The dictionary's key type is String and value is of a generic type: A.

Node insertion

Here is the first method, which inserts a new node, *persisting* the existing tree: dict. A new version of the dictionary is created and returned. The previous tree version will be garbage-collected if there are no references to it:

```
scala>    def insert[A](key: String, value: A, dict: Dictionary[A]):
Dictionary[A] = dict match {
     |       case Leaf => Branch((key, value), Leaf, Leaf)
     |       case Branch((k, v), l, r) if (k == key) => sys.error(s"key
${key} already present")
     |       case Branch((k, v), l, r) if (k > key) => Branch((k,v),
insert(key, value, l), r)
     |       case Branch((k, v), l, r) if (k < key) => Branch((k,v), l,
insert(key, value, r))
     |    }
```

The insert method tries to find an empty slot, a Leaf, in which to insert the new value. The method pattern matches the dictionary argument dict (which is just BinTree[A]).

The first clause matches when we hit a Leaf node:

```
case Leaf => Branch((key, value), Leaf, Leaf)
```

We have found an empty slot so we return Branch((key, value), Leaf, Leaf). Note that the Branch node's value is a tuple, holding the tuple (key, value) formed from the arguments.

The second case clause does a lot:

```
case Branch((k, v), l, r) if (k == key) => sys.error(s"key ${key} already
present")
```

It performs a *nested match*. The clause checks whether the node is `Branch`. It also extracts and binds the `value` tuple of `Branch`. There is a pattern guard too that compares the argument `key` with `k`, the `Branch` node's key.

As our dictionary holds unique keys, adding a duplicate is flagged as an error.

The next clause is as follows:

```
case Branch((k, v), l, r) if (k > key) => Branch((k,v), insert(key, value,
l), r)
```

This matches when the `Branch` node's key `k` is greater than the argument, namely `key`.

We copy the left subtree node to preserve the earlier version:

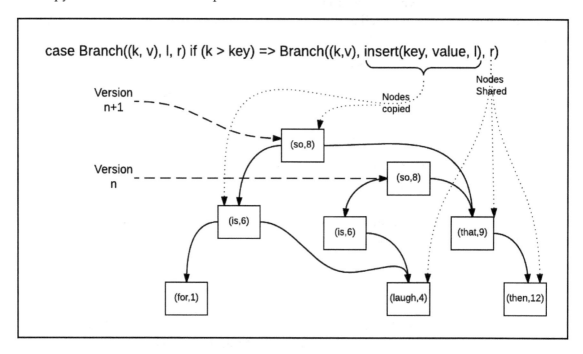

The root node, with the key `so`, and its left node, with the key `is`, are copied. The rest of the nodes, in this case all the right subtree nodes, are *structurally shared*.

The third clause is similar to this:

```
case Branch((k, v), l, r) if (k < key) => Branch((k,v), l, insert(key,
value, r))
```

The difference is that it copies the right subtree node, structurally sharing the left subtree.

Searching a key

Searching the dictionary is very simple. It looks up the `key` value and returns `Option[A]`. If `key` is absent, it returns `None`. Otherwise, it returns the value associated with the key, wrapped in `Some(value)`:

```
scala>  def search[A](key: String, dict: Dictionary[A]): Option[A] = dict
match {
       |       case Leaf => None
       |       case Branch((k, v), l, r) if (k == key) => Some(v)
       |       case Branch((k, v), l, r) if (k > key) => search(key, l)
       |       case Branch((k, v), l, r) if (k < key) => search(key, r)
       |    }
```

The analysis is straightforward and I've left it as an exercise.

Updating a value

The `update` method is similar to the `insert` method. If `key` is not present, the `(key, value)` pair is just inserted:

```
    def update[A](key: String, value: A, dict: Dictionary[A]): Dictionary[A]
= dict match {
    case Leaf => Branch((key, value), Leaf, Leaf)
    case Branch((k, v), l, r) if (k == key) => Branch((k, value), l, r)
    case Branch((k, v), l, r) if (k > key) => Branch((k, value),
update(key, value, l), r)
    case Branch((k, v), l, r) if (k < key) => Branch((k, value), l,
update(key, value, r))
    }
```

If `key` is found, `update` does not throw an error:

```
    case Branch((k, v), l, r) if (k == key) => Branch((k, value), l, r)
```

Instead, it replaces the value associated with the key. A new `Branch` node is created with the tuple value `(k, value)`, where `value` is the new argument value.

The rest of the method is similar to `insert` and tracing it out is left as an exercise.

Exercising it

It would be tedious to insert nodes manually to incrementally build a tree. Instead, we use *folding* to fold a list and *accumulate the BST* as a result.

We use the `foldLeft` idiom a lot to build other data structures too. Here is how we could sum up a `List[Int]`:

```
scala> val l = List(1,2,3,4)
l: List[Int] = List(1, 2, 3, 4)
scala> l.foldLeft(0)((acc, x) => acc + x)
res33: Int = 10
```

The method has a *curried form*. It takes an initial value 0 and a function. It keeps invoking the function *for each value of the list*. The *updated value of the accumulator* is passed for each function invocation.

Here's a quiz for you: Could you multiply the numbers instead of summing them up? Please see `https://coderwall.com/p/4173-a/scala-fold-foldleft-and-foldright` for more information:

```
scala>   def empty[A](): Dictionary[A] = Leaf
empty: [A]()Dictionary[A]
```

The `empty` method indicates an empty BST, which is just a `Leaf`. This forms our *initial value* for `foldLeft`:

```
scala>   val list = List(("so", 8),
     |       ("that", 9),
     |       ("is", 6),
     |       ("then", 12),
     |       ("laugh", 4),
     |       ("for", 12)
     |   )
list: List[(String, Int)] = List((so,8), (that,9), (is,6), (then,12),
(laugh,4), (for,12))
```

This is the tuple list, the input data for insertion. We invoke `foldLeft` on it:

```
scala>   val t = list.foldLeft(empty[Int]())((acc, x) => insert(x._1, x._2,
acc))
t: Dictionary[Int] =
Branch((so,8),Branch((is,6),Branch((for,12),Leaf,Leaf),Branch((laugh,4),Lea
f,Leaf)),Branch((that,9),Leaf,Branch((then,12),Leaf,Leaf)))
```

```
scala>    println(inorder(t))
List((for,12), (is,6), (laugh,4), (so,8), (that,9), (then,12))

scala>    println(preorder(t))
List((so,8), (is,6), (for,12), (laugh,4), (that,9), (then,12))

scala>    println(search("then", t))
12

scala>    println(search("when", t))
None
```

Using these methods, we could easily split a fairly long String and count the word frequencies. This is left as an exercise.

Summary

Trees are a hierarchical data structure with notions of *root*, *parent*, *children*, and *leaf* nodes. Binary trees are trees where each parent node can have a maximum of two children nodes, each of which could be a binary tree itself.

Binary trees are *recursively defined* data structures like List.

We defined the various types of nodes of a binary tree. We also implemented a number of related functional algorithms.

We used Binary Search Trees for implementing dictionaries. We learned about some Scala idioms too.

All of this know-how arms us well to look at more fascinating functional data structures. On to it!

5
More List Algorithms

In this chapter, we will look at some more list algorithms to see how lists permeate functional programming. For example, binary numbers could be represented as a list of 0 and 1.

We will apply all that we have learned so far to implement algorithms in order to perform binary arithmetic. Binary arithmetic works from right to left. The **least significant bit (LSB)** is the rightmost bit, where we will start working. However, we typically process lists from left to right. Getting a list head and prepending to a list are $O(1)$ operations. We will see how a change in representation allows us to design algorithms in terms of list head and list prepend.

While doing binary arithmetic, we need to deal with a carry. The carry operation is implemented as a helper method, which is reused as needed.

All these building blocks enable us to implement binary addition and multiplication.

We will wrap up with a look at greedy algorithms. We will look at a fun problem related to coins and use a greedy strategy to solve it.

When you are done with this chapter, you will have a very good grasp of how to apply lists for designing algorithms.

Binary numbers

We use a List[Int] to represent binary numbers, a list of 0's and 1's. If you pass in a list that has any other numbers except 0 or 1, the algorithms will throw an exception.

Before we look at the summation and multiplication operations, let's look at how to handle the carry operation. For example, when you add **1** to **1011** (decimal **11**), you get **1100** (decimal **12**). Here is how a carry is *propagated*:

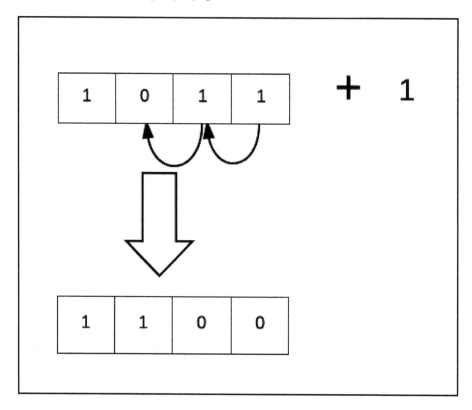

Before we try modeling the binary numbers as a list, there is a caveat we need to be aware of!

We write a binary number from left to right. In other words, the *most significant bit* is at the leftmost and the *least significant bit* of a binary number is at the rightmost.

To add a carry, we typically start from the right; however, as we have already seen, for a list, that would be the tail. Working on a list tail is expensive. Instead, we want to work at the *head* of the list and express operations using list *prepend*. This means we need to reverse the list so we can both work at the head of the list as well as start operating from the LSB bit onward.

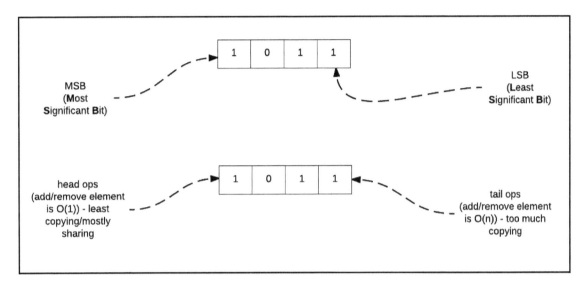

With this under our belt, here is the carry operation:

```scala
scala> def carry(c: Int, list: List[Int]):List[Int] = (c, list) match {
     |    case (0, xs) => xs
     |    case (1, Nil) => List(1)
     |    case (1, x :: xs) => (1 - x) :: carry(x, xs)
     |    case (_, _) => throw new IllegalArgumentException("Invalid
input!!!")
     | }
carry: (c: Int, list: List[Int])List[Int]
```

This is a helper function used in the summation function.

There are four match clauses. The first one is simple:

```scala
case (0, xs) => xs
```

When the carry bit is 0, there is no change and we return the list as is.

The next case is when the carry is 1 and the list empty:

```
case (1, Nil) => List(1)
```

The clause just returns a list that has 1.

The third clause is more interesting and illustrates a general case:

```
case (1, x :: xs) => (1 - x) :: carry(x, xs)
```

Here, we have the carry as 1 and a non-empty list. Let's say x is 1; if so, then we will have 0 and the carry will propagate to the rest of the list.

The final clause catches mistakes. For example, if we erroneously send a list with the digit 2, the clause catches it and aborts the program execution.

Here is how we exercise it:

```
scala> carry(1, List(1, 1, 0, 1))
res0: List[Int] = List(0, 0, 1, 1)
```

As noted, the number is 1011, and adding 1 to it, we get 1100. We work on it in reverse fashion so we can express operations on the LSB in terms of list head operations.

Addition

Let's try adding two binary numbers. For example, adding **01011** (that is 11) to **11110** (that is 30) should give us 41.

We use the preceding carry helper method to realize the addition:

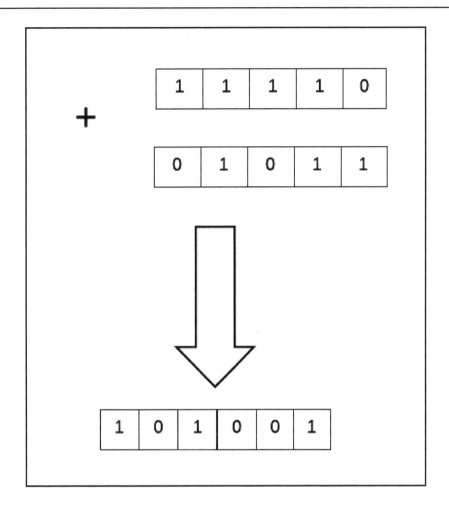

The preceding figure shows the addition of binary **11110** (decimal 30) to binary **01011** (decimal 11). Adding the second bits (from the right) generates a carry that keeps propagating. The addition results in 41.

Here is the add method:

```
scala> def add(c: Int, ps: List[Int], qs: List[Int]): List[Int] = (ps, qs)
match {
      |     case (Nil, Nil) => carry(c, Nil)
      |     case (Nil, _ :: _) => carry(c, qs)
      |     case (_ :: _, Nil) => carry(c, ps)
      |     case (x::xs, y::ys) => ((c+x+y) % 2) :: add((c+x+y)/2, xs, ys)
      | }
add: (c: Int, ps: List[Int], qs: List[Int])List[Int]
```

Let's look at this closely. We match the pair formed by both the lists. The carry bit is initially set to zero:

```
case (Nil, Nil) => carry(c, Nil)
```

If both the lists are empty and if the carry bit is 1, we return 1; if not, we return 0. Note how we push the latter part to the `carry` method:

```
case (Nil, _ :: _) => carry(c, qs)
```

This case is triggered when the first list is empty but the second is not. In this case, we just absorb the carry bit in the second list and return it.

Note the `_ :: _` part. We really don't care about binding the match to any identifier. We just need to make sure the list is non-empty. We use the underscore to stress that we don't care about the actual list elements:

```
case (_ :: _, Nil) => carry(c, ps)
```

This case is similar to the second case where the first list is not empty but the second one is:

```
case (x::xs, y::ys) => ((c+x+y) % 2) :: add((c+x+y)/2, xs, ys)
```

This is the general case where both the lists are not empty.

Let's say we have $c = 1$, $x = 1$, and $y = 1$. The addition leaves 1 behind and the carry becomes 1. We propagate the carry along to the rest of the addition.

Working out the rest of the cases is left for you as an exercise. We could improve the interface a bit, as sending an initial 0 for a carry is really part of implementation detail.

Here is the `addNums` function, which in turn calls `add`:

```
scala> def addNums(first: List[Int], second: List[Int]): List[Int] = {
     |    val result = add(0, first.reverse, second.reverse)
     |    result.reverse
     | }
addNums: (first: List[Int], second: List[Int])List[Int]
```

We do better with `addNums`; it hides all the list reversals so the usage feels more natural:

```
scala> addNums(List(1, 0), List(1, 0))
res1: List[Int] = List(1, 0, 0)

scala> addNums(List(1, 0, 1, 1), List(1, 1, 1, 1, 0))
res2: List[Int] = List(1, 0, 1, 0, 0, 1)
```

Try out some more examples to get a feel of it.

Multiplication

Multiplying two binary numbers is just a little more elaborate than the process of adding them. Multiplying is repeated addition! For example, multiplying **1111** (decimal 15) with **11** (decimal 3) gives us **101101** (decimal 45). We just add `1111` to itself thrice.

The following figure shows the actual operation performed:

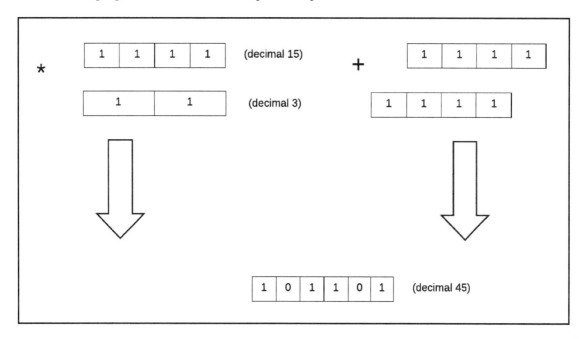

Here, both the operands are bits and they are 1. What happens when there are 0 bits?

For each 0 bit, we shift to the left. It might help to remember that when we multiply a number by two (binary 10), we do a bitwise shift left. (See `http://stackoverflow.com/questions/6385792/what-does-a-bitwise-shift-left-or-right-do-and-what-is-it-used-for` for an overview of bitwise shifting.)

The following diagram depicts the process when we multiply the binary **111** (decimal 7) with the binary **101** (decimal 5). The result is 35.

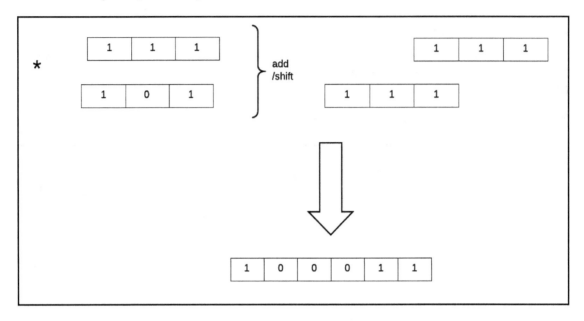

Here is the code; we hide the reversal process using an inner method called `multiply`:

```
scala> def mult(first: List[Int], second: List[Int]): List[Int] = {
     |     def multiply(ps: List[Int], qs: List[Int]): List[Int] = (ps) match
{
     |         case Nil => Nil
     |         case 0 :: xs => 0 :: multiply(xs, qs)
     |         case 1 :: xs => add(0, qs, 0::multiply(xs, qs))
     |     }
     |     val result = multiply(first.reverse, second.reverse)
     |     result.reverse
     | }
```

As shown, the `mult` method takes two lists, each made up of 0 and 1.

The first clause is simple:

```
case Nil => Nil
```

If the left operand is empty, we return an empty list. There is just nothing left to multiply with.

The second case is when the current bit is 0. This does a bitwise left shift:

```
case 0 :: xs => 0 :: multiply(xs, qs)
```

The following diagram shows the case when the current bit is 1:

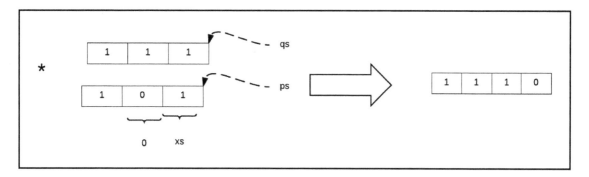

The third case is when the current bit is 1:

```
case 1 :: xs => add(0, qs, 0::multiply(xs, qs))
```

In this case, we just add qs to the result of multiplying xs and qs. This is shown in the following figure:

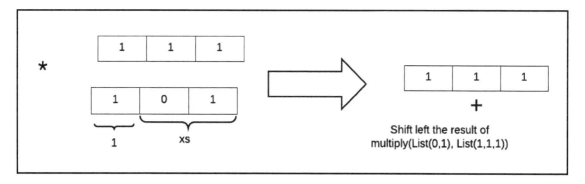

Here's an exercise for you:

Fix the match error in the mult method.

Greedy algorithms and backtracking

What do we mean by greedy algorithms? What is backtracking? By being *greedy*, the algorithm matches the longest possible part. *Backtracking* algorithms, upon failure, keep exploring other possibilities. Such algorithms begin afresh from where they had originally started, hence they *backtrack* (go back to the starting point).

We all follow the process of backtracking in real life. For example, to get to an address, we go to a well-known landmark, then try the first lane, for example. If there is no success, we backtrack to the landmark again and try another lane (we may ask a passerby for help). We keep doing this until we get to the address or give up the search altogether.

A well-known example of greedy and backtracking algorithms from the programming world is how a regex match happens. We use a simple regex using greedy qualifiers such as * and +:

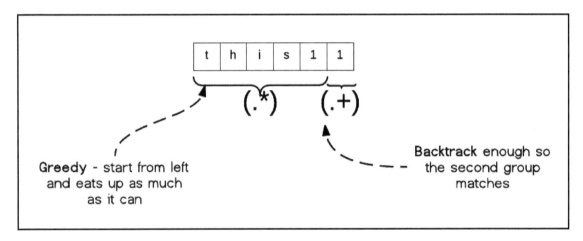

Here is a quick Scala REPL session to see the greediness in action:

```scala
scala> val regex = "^(.*)(.+)$".r
regex: scala.util.matching.Regex = ^(.*)(.+)

scala> val regex(first, second) = "this11"
first: String = this1
second: String = 1
```

We define a regex by calling the `.r` method of `StringOps`. A string has an implicit conversion to `StringOps`. So the `"^(.*)(.+)"` string gets converted into `StringOps`; then its `.r` method is called that finally creates the regex.

The second line `val regex(first, second) = "HowDoYouDo"` uses an *extractor*. Similar to a pattern match, the extractor binds the variables first and then to the matching groups.

See *Chapter 26, Extractors in Programming in Scala, Programming in Scala, Third Edition*, for a detailed explanation of various extractors. For more information please refer to `http://www.journaldev.com/8270/scala-extractors-apply-unapply-and-pattern-matching`.

More information on the basics of regex matching and the *leftmost longest rule* are available at `http://www.regular-expressions.info/repeat.html`.

The greediness is evident in *how the first matching group is captured*. The regex matching rule is *leftmost, longest*–meaning, the matching starts from the *leftmost* part and is *greedy*. It gobbles up as many characters as it can and *only stops* when there is a danger of *failing the overall match*.

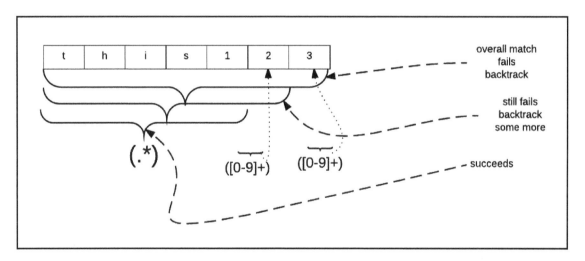

Here is the REPL snippet to see the backtracking in action:

```scala
scala> val regex = "^(.*)([0-9]+)([0-9]+)$".r
regex: scala.util.matching.Regex = ^(.*)([0-9]+)([0-9]+)$

scala> val regex(first, second, third) = "HowDoYouDo1234"
first: String = HowDoYouDo12
second: String = 3
third: String = 4
```

As noted, the group captured in the `first` variable is greedy, so it has the trailing 12 too. The variables `second` and `third` just get a digit each, namely 3 and 4.

An example of a greedy algorithm

We now apply what we have learned to a fun problem–given a coin set, how could we tender change such that a certain amount is made? You could use the same coin again and again.

For example, given a set of coins of denomination (7,2), an amount of 16 would mean (7,7,2):

```scala
scala> greedyChange(List(7, 2), 16)
res23: List[Int] = List(7, 7, 2)
scala> greedyChange(List(7, 2), 20)
res25: List[Int] = List(7, 7, 2, 2, 2)
```

The first argument is the list of coin denominations, sorted in decreasing order. The second argument is the amount we need to make up.

Here is the greedy algorithm:

```scala
scala> def greedyChange(denom: List[Int], amount: Int): List[Int] = (denom,
amount) match {
     |    case (_, 0) => Nil
     |    case (x :: xs, _) if (amount < x) => greedyChange(xs, amount)
     |    case (x :: xs, _) => x :: greedyChange(denom, amount - x)
     | }
greedyChange: (denom: List[Int], amount: Int)List[Int]
```

Here is how it works:

```scala
case (_, 0) => Nil
```

This first clause is simple. When we ask for an amount totaling 0, we don't even have to look at the coin denominations. We just return `Nil` as there is nothing to cover:

```scala
case (x :: xs, _) if (amount < x) => greedyChange(xs, amount)
```

The second clause is when the amount is less than the current coin denomination value. Remember, the coin denominations are in *decreasing order*. So we just move on to the next (lesser) value and try again:

```scala
case (x :: xs, _) => x :: greedyChange(denom, amount - x)
```

The third clause is when the current coin value is *less than the amount*. So we deduct the largest value from the amount and repeat the process.

This operation is *greedy* as we always try deducting the largest value. This is similar to the greedy regex matching we just saw.

Here is a pictorial version of the running algorithm:

There is a problem with the greedy version though. Consider the following call:

```
scala> greedyChange(List(5, 2), 16)
scala.MatchError: (List(),1) (of class scala.Tuple2)
    at .greedyChange(<console>:11)
```

However, we could very well pay 16 as (5, 5, 2, 2, 2)! The problem is that *greediness* eats too much of the change. It doesn't stop until 15; then, for the pending 1, it has no coin denomination left.

The backtracking jig

We now look at the backtracking version of the algorithm. We show a couple of examples first; the implementation follows just after.

Here is how the backtracking algorithm handles the previously failing case:

```
scala> btChanges(List(), List(5,2), 16)
res42: List[List[Int]] = List(List(2, 2, 2, 5, 5), List(2, 2, 2, 2, 2, 2,
2, 2))
```

It works–we also get *all the possible* distributions! Try adding a coin with value 1 to the denominations:

```
scala>  btChanges(List(), List(5,2,1), 16)
res49: List[List[Int]] = List(List(1, 5, 5, 5), List(2, 2, 2, 5, 5),
List(1, 1, 2, 2, 5, 5), List(1, 1, 1, 1, 2, 5, 5), List(1, 1, 1, 1, 1, 1,
5, 5), List(1, 2, 2, 2, 2, 2, 5), List(1, 1, 1, 2, 2, 2, 2, 5), List(1, 1,
1, 1, 1, 2, 2, 2, 5),...
```

Here comes the implementation: you need to enter the REPL's paste mode to enter the code. Enter the paste mode by typing :paste at the REPL prompt. Exit it by entering *Ctrl + D*:

```
scala> :paste
// Entering paste mode (ctrl-D to finish)

def btChanges(result: List[Int], denom: List[Int], amount: Int):
  List[List[Int]] =
  if (amount == 0)
    List(result)
  else
    denom match {
  case Nil => List()
  case x :: xs if amount < 0 => List()
  case x :: xs => btChanges(x::result, denom, amount - x) ++
btChanges(result, xs, amount)
}

// Press Ctrl-D
// Exiting paste mode, now interpreting.

btChanges: (result: List[Int], denom: List[Int], amount:
Int)List[List[Int]]
```

We make an additional change, passing in the result list explicitly:

```
if (amount == 0)
   List(result)
```

If the amount is 0, we just return the current result list. Otherwise, we do a pattern match on the coin's denom list:

```
case Nil => List()
```

If the `denom` list is empty, that clearly means we have an amount *that is not zero and there are no more coins left*:

```
case x :: xs => btChanges(x::result, denom, amount - x) ++
btChanges(result, xs, amount)
```

This is a very interesting clause–it has two recursive calls. Note that to *make progress, we have to reduce something*! In the first call, the amount varies:

```
btChanges(x::result, denom, amount - x)
```

This is the first call in which the amount is what remains after we subtract the current coin:

```
btChanges(result, xs, amount)
```

The second call is where the denomination list varies, whereas the amount remains the same.

Here is a part of the execution depicted pictorially. Completing the diagram is left as an instructive exercise for you to understand the flow:

Note how we enumerate all the possible paths and discard invalid solutions and collect just the right ones.

Another important aspect is that the execution always varies, either in terms of the amount or the denomination of the coins. This is essential so that we keep reducing possibilities, thereby guaranteeing eventual termination.

Trace out a couple of examples and you will appreciate the beauty of the algorithm. Expressing it using the functional paradigm yields succinct and expressive code.

Summary

We looked at some more list algorithms in this chapter and how lists permeate functional programming.

Modeling binary numbers as an integer list of just 0 or 1 is very easy. List operations perform best when they work at the head, allowing you to have maximum sharing and least copying. So we reversed the number's list to express our algorithms.

We looked at addition and multiplication algorithms. Next, we covered the concept of greediness and backtracking. We saw how these concepts drive regular expression matching.

We then looked at a fun problem: how to satisfy an amount, given a set of coin denominations. We looked at the greedy version, which proved to be too greedy at times. This was fixed with a backtracking solution.

Hopefully, this gives you enough ground to look at more exciting functional algorithms. Let's continue the exciting journey!

6
Graph Algorithms

How does immutability affect algorithm design? How are typical algorithms implemented without resorting to in-place mutation?

This chapter will give you a taste of functional algorithms. List prepending will be one dominating theme here. We will start by looking at list reversal and how prepending helps in dealing with algorithms.. We will then look at an efficient algorithm for list reversal using list prepending.

Graphs are a very important data structure; they are used to model related entities. We will be looking at directed graphs, also known as **digraphs**. We will implement functional versions of common digraph algorithms, for example, traversing a graph and visiting each node to do something useful.

Topological sorting is a digraph algorithm used to compute nodes' order of precedence. Build tools, such as Make, need to order tasks based on precedence. Topological sorting is the algorithm used by such tools to process a task's dependencies before the actual task is processed.

We will implement functional topological sorting and improve it to handle digraphs with cycles.

By the end of this chapter, you will have seen some design principles for a functional algorithm design.

Reversing a list

Let's look at how to reverse a list. Here is our first implementation:

```scala
scala> def slowRev(list: List[Int]): List[Int] = list match {
     |    case Nil => Nil
     |    case x :: xs => slowRev(xs) :+ x
     | }
slowRev: (list: List[Int])List[Int]

scala> slowRev(List(1,2,3))
res2: List[Int] = List(3, 2, 1)
```

This is not a very efficient algorithm. We end up doing too much copying. The following diagram shows the copying of nodes for a list with three elements:

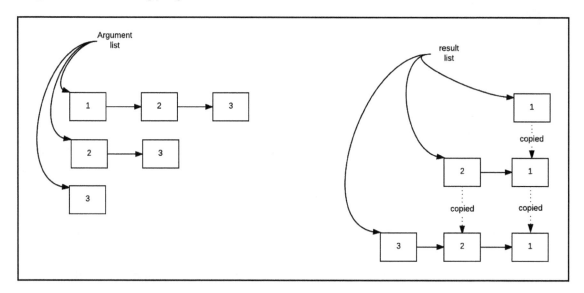

As we know from Chapter 2, *Building Blocks*, every time we append an element, we need to *copy* the entire source list. We cannot resort to pointer juggling, as *in-place update* is not allowed. Never forget that we work with immutable and persistent data structures.

An append operation has $O(n)$ complexity. As there are n appends, the total complexity is $O(n^2)$.

The preceding version is not tail-recursive and hence is prone to Stack Overflow errors:

```scala
scala> import scala.annotation.tailrec
import scala.annotation.tailrec

scala> @tailrec
     | def rev(list: List[Int], acc: List[Int]): List[Int] = list match {
     |   case Nil => acc
     |   case x :: xs => rev(xs, x :: acc)
     | }
rev: (list: List[Int], acc: List[Int])List[Int]

scala> rev(List(1,2,3), Nil)
res1: List[Int] = List(3, 2, 1)
```

The complexity of this method is clearly *O(n)*. Instead of copying nodes, we *share* them among multiple versions of the list.

The following diagram shows this algorithm in action:

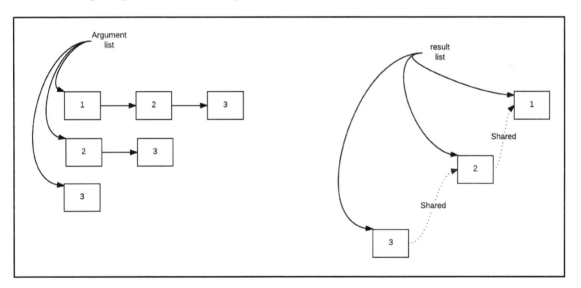

Maximizing sharing and minimizing copying are fundamental techniques when we design list-based functional algorithms.

Let's look at some more list-based algorithms and see these techniques in action.

Graph algorithms

A graph has a finite number of *nodes*, also commonly called **vertices**, connected via *edges*. Trees are special cases of graphs, with additional constraints.

There are undirected and directed graphs. We will be looking at directed graphs, also known as digraphs. I will explain the terms as they come along; however, `http://algs4.cs.princeton.edu/42digraph/` would help as a quick refresher/introduction.

A list of pairs will be used to model these directed graphs. The second element of the pair is should be a *successor* of the first. For example, the m node's successors are {n,p,o}:

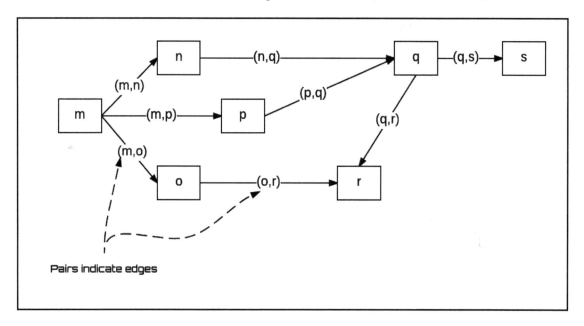

Pairs indicate edges

The preceding graph is modeled in the following code. Each edge is denoted by a pair, and the graph is a list of such pairs:

```
scala> val graph = List(("m", "n"), ("m", "o"), ("m", "p"),
     |                  ("n", "q"), ("o", "r"), ("p", "q"),
     |                  ("q", "r"), ("q", "s"))
graph: List[(String, String)] = List((m,n), (m,o), (m,p), (n,q), (o,r),
(p,q), (q,r), (q,s))
```

The `succSet` method collects the set of successors of a given vertex:

```scala
scala> def succSet(a: String, g: List[(String, String)]): List[String] =
g match {
     |    case Nil => Nil
     |    case x :: xs if (a == x._1) => x._2 :: succSet(a, xs)
     |    case _ :: xs => succSet(a, xs)
     | }
succSet: (a: String, g: List[(String, String)])List[String
```

Let's try it out:

```scala
scala> succSet("m", graph)
res1: List[String] = List(n, o, p)
```

This is a pretty simple method. We traverse the graph and check the first element; if it equals the argument vertex, we collect the second element, which is the successor.

Here is an idiomatic way to write the preceding method:

```scala
scala> graph filter (_._1 == "m")
res3: List[(String, String)] = List((m,n), (m,o), (m,p))

scala> graph filter (_._1 == "m") map (_._2)
res4: List[String] = List(n, o, p)
```

This method is an important building block for the upcoming algorithms, as you will soon see.

Graph traversal

Graph traversal is the process of visiting each node in the graph. There are two strategies: either a *depth-first traversal* or *breadth-first traversal*. The name depth-first indicates that a node's children are visited *before* its siblings. On the other hand, in a breadth-first strategy, all the node's siblings are visited first and then its children.

In either traversal, we visit a node only once.

We will be implementing a depth-first traversal. Implementing the breadth-first traversal is left as an exercise for you. Now check out the following figure:

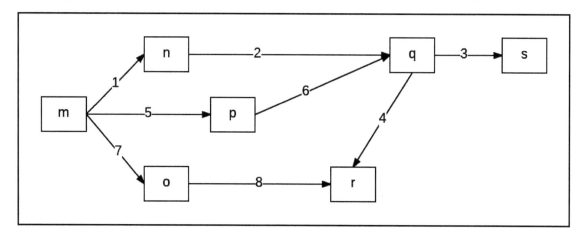

As shown in the diagram, we start the traversal at node m. This node has three children (its successors): {n, p, o}.

We visit the node n first and then the n node's child, namely q.

The traversal always visits children first, so when we are at q, we go to s and r (the other way, that is, visiting r and then s would also work).

As r and s have no children, we return to q, then to n, and finally to m. The traversal next visits p, which is the second child of m. Now when we start looking at the p node's only child, that is q, we find that it was visited earlier in the traversal. So the q node is skipped.

Lastly, o is visited, and when at o, we find its only child r has already been visited. So it is skipped.

This completes our traversal.

Here is the traversal algorithm code:

```
scala> def depthFirst(initial: String, g: List[(String, String)]) :
List[String] = {
     |    def depthf(nodes: List[String], visited: List[String]):
List[String] = nodes match {
     |       case Nil => visited
     |       case x :: xs if visited.contains(x) => depthf(xs, visited)
     |       case x :: xs => depthf(succSet(x, g) ++ xs, x::visited)
     |    }
```

```
    |
    |    val result = depthf(List(initial), List())
    |    result.reverse
    | }
depthFirst: (initial: String, g: List[(String, String)])List[String]

scala> depthFirst("m", graph)
res5: List[String] = List(m, n, q, r, s, o, p)
```

The method implements the algorithm outlined in the preceding code. We use an internal helper method: `depthf`. This method is the real workhorse function.

Note that the input graph is never modified; however, it needs to be referenced. We could have written just one method, but it would require passing the graph as an argument.

Is there a way we could *stay away from passing the same argument*, but at the same time *have it in the scope so it could be referred to*? Closures do just this!

Here is an example:

```
scala> var m = 10
m: Int = 10
scala> def callAndPrint(x: Int, f: Int => String) =
    |    println(s"<<${f(x)}>>")
callAndPrint: (x: Int, f: Int => String)Unit
scala> callAndPrint(5, (t: Int) => (t+m).toString)
<<15>>
scala> callAndPrint(5, (t: Int) => (t-m).toString)
<<-5>>
```

We define a variable, for example, m. The `callAndPrint` method takes a function as an argument. It calls the function, passing in the first argument, and prints the result.

Note the argument `(t: Int) => (t+m).toString`, which is an unnamed function literal. For the function literal, the variable m is in scope. As far as the literal is concerned, m is a free variable and t is a bound variable.

This explains why the variable g is in scope for the `depthf` method. We prefer to keep the `depthf` method private, thereby hide its implementation details:

```
def depthf(nodes: List[String], visited: List[String]): List[String]
```

We use `visited`, a list of nodes, to keep track of already visited nodes. Note that we use list prepending to add the node x to the `visited` list.

However, the visited nodes are saved in *reverse* order of visitation. We can fix this in a minute though:

```
nodes match {
|      case Nil => visited
```

This case clause is hit when we have no more edges to traverse. In other words, reaching here means the traversal is complete. We just return the `visited` list:

```
case x :: xs if visited.contains(x) => depthf(xs, visited)
```

This clause matches when x is a member of the `visited` list. This tells us to skip x as it was visited already in an earlier traversal and got added to `visited`.

So we just continue our traversal with the remaining node elements:

```
case x :: xs => depthf(succSet(x, g) ++ xs, x::visited)
```

This is an interesting clause, which is matched when we encounter the node x for the first time.

We take the children of x via a call to the `succSet` method. We prepend this list so the children are visited before `xs`, the siblings. This makes the traversal *depth-first*.

Here is a refresher, showing the `List` method ++ in action:

```
scala> List(1,2,3) ++ List(4, 5, 6)
res8: List[Int] = List(1, 2, 3, 4, 5, 6)
```

The first list of elements appears before the second list elements.

The enclosing function `depthFirst` is just a shell. Its job is to present a convenient public interface to the caller:

```
|      val result = depthf(List(initial), List())
|      result.reverse
```

It converts the initial traversal node into a list with one element. It sends an empty list for the second argument of the `depthf` method. It also reverses the result, correcting the reversal introduced by the prepend operation.

Avoiding list appending

In Chapter 2, *Building Blocks*, we saw how adding two lists works.

Here is a diagram showing list additions:

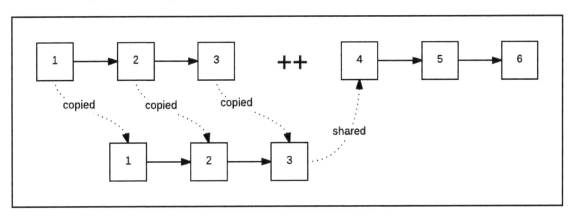

The first list is copied and the second is structurally shared. For every node's children list, we pay for this copying.

Could we do better, thereby avoid this expensive list addition? Check out the following code:

```
scala> def depthFirst1(initial: String, g: List[(String, String)]) :
List[String] = {
     |    def depthf(nodes: List[String], visited: List[String]):
List[String] = nodes match {
     |       case Nil => visited
     |       case x :: xs => depthf(xs,
     |         if (visited.contains(x)) visited
     |         else depthf(succSet(x, g), x::visited))
     |    }
     |
     |    val result = depthf(List(initial), List())
     |    result.reverse
     | }
depthFirst1: (initial: String, g: List[(String, String)])List[String]
```

This slightly complicated version avoids the list append. The only change is how we handle the case of visiting a node for the first time:

```
|        case x :: xs => depthf(xs,
|           if (visited.contains(x))  visited
|           else depthf(succSet(x, g), x::visited))
|     }
```

We make use of Scala's if...else statement, which is an expression:

```
scala> val x = 10
x: Int = 10
scala> val k = if (x > 5) 20 else 30
k: Int = 20
scala> val k = if (x < 5) 20 else 30
k: Int = 30
```

The point to note is that if the condition is true, the "if" body is evaluated and its value is the result of the entire expression. If the condition is false, the body of the "else" clause is evaluated and assigned likewise.

We use this feature and express the algorithm differently:

```
else depthf(succSet(x, g), x::visited
```

We go and visit the children of *xright away* and then continue with the earlier nodes, thereby avoid having combination of any list!

Topological sorting

Topological sorting is a sorting algorithm for sorting tasks as per precedence. Certain tasks, for example, need to happen *before* other *dependent tasks*. The *compile* step needs to happen before we *run* the code. *Saving* the source code needs to happen before we take the compilation step. We usually shower before we dress, and not the other way round.

Task precedence is central to build tools such as *make* or *maven*. For example, you can't run tests after you have deployed the app.

Here is the routine of a typical office day as well as an off day:

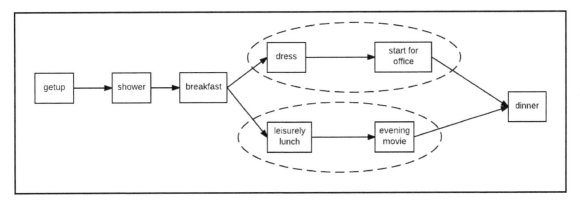

There are two linear sequences of events here. The first shows chores of a typical office day:

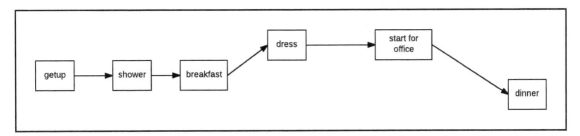

The second shows the events on an off day:

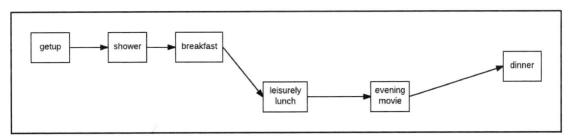

Finding such linear sequences in a graph is **topological sorting**.

Here is a code for the graph defining the preceding activities:

```
scala> val grwork = List(("getup","shower"),
     |    ("shower", "breakfast"),
     |    ("breakfast","dress"),
     |    ("dress","office"),
     |    ("office", "dinner"),
     |
     |    ("breakfast","leisurely_lunch"),
     |    ("leisurely_lunch", "movie"),
     |    ("movie", "dinner"))
```

With a small change to the depthFirst1 method, we get the sorting:

```
scala> def topsort(g: List[(String, String)]) = {
     |    def sort(nodes: List[String], visited: List[String]): List[String]
 = nodes match {
     |        case Nil => visited
     |        case x :: xs => sort(xs,
     |          if (visited.contains(x)) visited
     |          else x :: sort(succSet(x, g), visited))
     |    }
     |
     |    val (start, _) = g.unzip
     |    val result = sort(start, List())
     |    result
     | }
```

What is the change? The current node x is just *prepended* to the *recursive result* of topologically sorting *its successors*:

```
     |        case x :: xs => sort(xs,
     |          if (visited.contains(x)) visited
     |          else x :: sort(succSet(x, g), visited))
```

If x is already visited, the case is as before; we just skip it. If x is seen for the first time, we (topologically) sort its children and prepend x to the result.

Here it is in action:

```
scala> topsort(grwork)
res0: List[String] = List(getup, shower, breakfast, leisurely_lunch, movie,
dress, office, dinner)
```

The task *precedence* is correctly computed. Here is another run with the graph (which is just a list of pairs) *reversed*:

```scala
scala> topsort(grwork.reverse)
res1: List[String] = List(getup, shower, breakfast, dress, office,
leisurely_lunch, movie, dinner)
```

Again, task precedence is correctly computed.

Cycle detection

What if we introduce a cycle? For example, refer to the cycle shown here:

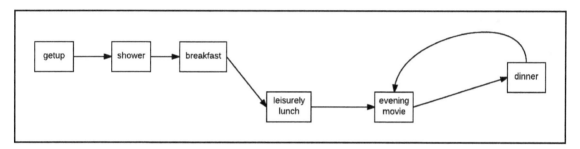

If you invoke the `topsort` method with this graph as an argument, it will never be completed:

```scala
scala> topsort(("dinner", "movie") :: grwork)
java.lang.StackOverflowError
...
```

The fix is to remember that x is already `seen`, going to `precede`, so processing x again is a cycle. We just abort the execution in this case:

```scala
scala> def topsortWithCycle(g: List[(String, String)]) = {
    |    def sort(nodes: List[String], path: List[String], visited:
List[String]):
    |    List[String] = nodes match {
    |      case Nil => visited
    |      case x :: xs if path.contains(x) =>
    |        throw new RuntimeException("Cycle detected")
    |      case x :: xs => sort(xs, path,
    |        if (visited.contains(x)) visited
    |        else x :: sort(succSet(x, g), x :: path, visited))
    |    }
    |
```

```
        |    val (start, _) = g.unzip
        |    val result = sort(start, List(), List())
        |    result
        | }
topsortWithCycle: (g: List[(String, String)])List[String]

scala> topsortWithCycle(("dinner", "movie") :: grwork)
java.lang.RuntimeException: Cycle detected
```

The path list stores x and aborts it if we attempt to add x again.

We could instead return an empty list if a cycle is detected. This is left as an exercise for you.

Printing the cycle

Instead of throwing an exception and aborting the program execution, we could print the cycle. We will return a pair of lists. The first list would contain topologically sorted nodes, whereas the second would contain the offending nodes, creating cycles.

To make things simpler, we can resort to Scala's type keyword to create type alias:

```
scala> type VC = (List[String], List[String])
defined type alias VC
```

So, instead of writing the longer type declaration, we could write a shorter version:

```
scala> val v1 : (List[String], List[String]) = (Nil, Nil)
v1: (List[String], List[String]) = (List(),List())

scala> val v1 : VC = (Nil, Nil)
v1: VC = (List(),List())
```

This makes our code much more readable and easier to understand. We use this type alias to define a helper method and then the main sorting method.

First, the helper method:

```
scala> def addToVisited(x: String, v: VC) = (x :: v._1, v._2)
addToVisited: (x: String, v: VC)(List[String], List[String])

scala> addToVisited("h", (List("a", "b"), List()))
res7: (List[String], List[String]) = (List(h, a, b),List())
```

This method takes a string and adds it to the first member of the pair, a list. In the following method, this is the list that will contain the sorted nodes:

```scala
scala> def topsortPrintCycle(g: List[(String, String)]) = {
     |    def sort(nodes: List[String], path: List[String],
     |            visited: VC): VC = nodes match {
     |      case Nil => visited
     |      case x :: xs =>
     |        val (v, c) = visited
     |        sort(xs, path,
     |          if (path.contains(x)) (v, x::c)
     |          else if (v.contains(x)) visited
     |          else addToVisited(x, sort(succSet(x, g), x :: path,
visited))
     |        )
     |    }
     |
     |    val (start, _) = g.unzip
     |    val result = sort(start, List(), (List(), List()))
     |    result
     | }
topsortPrintCycle: (g: List[(String, String)])VC
```

Here is the code exercising the method:

```scala
scala> topsortPrintCycle(("dinner", "movie") :: grwork)
res5: VC = (List(getup, shower, breakfast, leisurely_lunch, dress, office,
dinner, movie),List(dinner))
```

As you can see, we got the topologically sorted nodes in the first list. The second list contained `dinner`, the node at which there is a cycle.

Summary

We looked at lists again, but in a different light. We revisited the list prepending and appending techniques and saw that prepending a node to a list has *O(1)* complexity. It is a very fast operation, sharing most of the existing list.

Appending to a list is very costly though, as we end up copying the entire existing list. We looked at list reversal and saw how we could express the list reversal algorithm in terms of list prepending.

Next, we saw how extensively list prepending is used. We looked at directed graphs, modeling them as a list of pairs.

We also implemented common graph algorithms, such as getting successors of a node and a depth-first traversal.

Incrementally tweaking the depth-first traversal, we came up with topological sorting, a sequence that respects precedence. We also implemented cycle detection and printing.

Hopefully, this gave you a taste of functional algorithms. In the next chapter, we will look at random access lists, yet another fascinating data structure–stay tuned!

7
Random Access Lists

Lists are great when we are prepending or matching at the head, having $O(1)$ complexity. However, as we saw, lists don't perform well when it comes to random element access. Accessing an element at the nth index has $O(n)$ complexity.

Starting at the head, we have to traverse (and skip) all the intervening list elements until we reach the nth element.

Arrays are another fundamental data structure; they allow you to have random access to any element without incurring any additional runtime cost.

Can we tweak our list implementations so random access to elements could be faster? In this chapter, we will see a list implementation that provides efficient lookup and update operations in addition to the usual head, tail, and cons operations.

Understanding the binary numerical representation is the key.

Earlier, we saw how we could model binary numbers as lists. We looked at the addition and multiplication of such lists. We will briefly look at the binary operations again, but in a different light. First, we will look at how to increment a binary number. Next, we will learn how to add two binary numbers. We will use the insight thus obtained to design a random access list, a list of binary trees. Note that all the binary tree roots are linked.

We will look at the element insertion, lookup and update algorithms. We will see how all these operations translate into equivalent binary tree operations. This is the key for improving the performance from $O(n)$ to $O(logn)$.

Incrementing a binary number

Before we jump into list implementation, let's look at the related binary operations.

Here is how we would increment a binary number represented by a list of 0 and 1:

```scala
scala> def increment(numList: List[Int]): List[Int] = numList match {
     |     case Nil => List(1)
     |     case 0 :: xs => 1 :: xs
     |     case 1 :: xs => 0 :: increment(xs)
     |     case _ => sys.error("Not a binary number")
     | }

scala> increment(List(1,0,1))
res3: List[Int] = List(0, 1, 1)

scala> increment(List(0, 1))
res4: List[Int] = List(1, 1)

scala> increment(List(0, 0, 1))
res5: List[Int] = List(1, 0, 1)
```

 Note that we need to write the numbers such that the *least significant bit* is on the *left-hand side*.

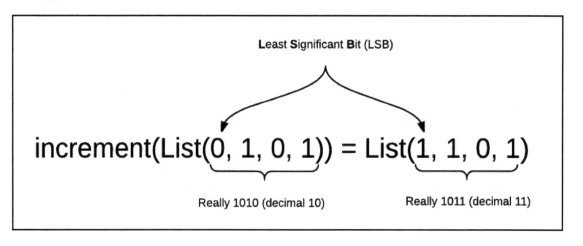

When we wrote the binary number algorithms, we saw the reason for it. We wanted the LSB to appear at the head of the list. Operating at the head of the list is an *O(1)* operation.

The `increment` method has four case clauses. The first one, and the simplest, is when we increment an empty list:

```
case Nil => List(1)
```

We define this operation to yield `List(1)`.

The second clause is as follows:

```
case 0 :: xs => 1 :: xs
```

This adds `1` to a number that ends in `0`. Remember the need for reversal.

For example, consider we're adding 1 to any binary number that ends in 0, such as 1000 (decimal 8). Here, we just replace the LSB 0 with 1, yielding 1001 (decimal 9).

The third case is when the LSB is `1`:

```
case 1 :: xs => 0 :: increment(xs)
```

Adding `1` to `1` results in `0` and a carry. So we recursively call the increment function again with the rest of the list to account for the carry.

 A carry is really adding 1 to the rest of the list.

The last clause is a *catch all*. It is an error to pass anything other than a sequence of `0` and `1`.

Here is a quick use of the method in REPL:

```
scala> increment(List(1, 0, 1))
res0: List[Int] = List(0, 1, 1)
```

Adding two binary numbers

Now let's look at how to add two binary numbers:

```
scala> def add(one: List[Int], two: List[Int]): List[Int] = (one, two)
match {
     |    case (Nil, Nil) => Nil
     |    case (xs, Nil) => xs
     |    case (Nil, xs) => xs
     |    case (x :: xs, 0 :: ys) => x :: add(xs, ys)
     |    case (0 :: xs, y :: ys) => y :: add(xs, ys)
     |    case (1 :: xs, 1 :: ys) => 0 :: increment(add(xs, ys))
     |    case _ => sys.error("Not a binary number")
     | }
```

The first three clauses are simple. If one of the lists is `Nil`, we return the rest. If both are `Nil`, we return `Nil`:

```
case (x :: xs, 0 :: ys) => x :: add(xs, ys)
```

This clause matches when both the lists are non-empty and the second LSB is 0. In this case, we just tack on the LSB of the first list (which could be either 0 or 1) and add the rest of the numbers:

```
case (0 :: xs, y :: ys) => y :: add(xs, ys)
```

This is the same as the fourth clause from before; the difference is just that the LSB of the first number is 0. It is handled in exactly the same way:

```
case (1 :: xs, 1 :: ys) => 0 :: increment(add(xs, ys))
```

This is the final clause, which is hit when both the LSBs are 1. As we know, this results in a carry:

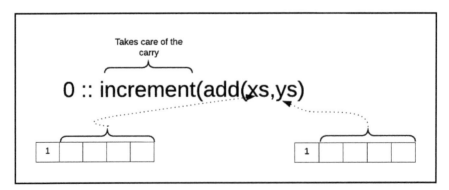

We take 0 and add a carry to the rest of the numbers' addition by incrementing the result. Here is how you can use the addition method:

```
scala> add(List(1, 1), List(1, 1))
res2: List[Int] = List(0, 1, 1)
```

We are adding 11 (decimal 3) to another 11 (decimal 3 again). The result is 011 which when reversed is 110; this means we have decimal 6.

List of tree roots

Time to change gears; we know that binary trees' lookup and update operations are fast. *Fast lookup* is the result of storing actual data elements in binary trees, that is, all the roots linked up in a list.

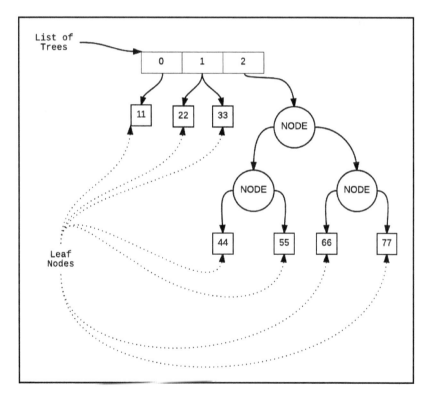

Here are our tree definitions:

```
scala> sealed abstract class Tree {
     |    def size: Int
     | }
```

Note the use of the `sealed` keyword. It is sealed, which means the definition could be used in this file only. We have an abstract method to know how many data elements (leaves) exist in this tree:

```
scala> case class Leaf(n: Int) extends Tree {
     |    override def size = 1
     | }
```

The `Leaf` class actually holds the data items. As the name indicates, it is a leaf and there are no children under a leaf:

```
scala> case object Zero extends Tree {
     |    override def size = 0
     | }
```

The `Zero` is a singleton object; it corresponds to the binary 0. We will see it in action pretty soon:

```
scala> case class One(t: Tree) extends Tree {
     |    override def size = t.size
     | }
```

The `One` class holds a tree; it corresponds to the binary 1. Just like `Zero`, we will soon see how it works:

```
scala> case class Node(sz: Int, left: Tree, right: Tree) extends Tree {
     |    override def size = sz
     | }
```

A `Node` is an internal node of the tree. It does not hold data; instead, it holds a tree.

Insertion

We have a helper method called `link` to link two trees. This helper is used by the `addTreeToList` method, which we will soon see:

```scala
scala> def link(t1: Tree, t2: Tree) = Node(t1.size + t2.size, t1, t2)
link: (t1: Tree, t2: Tree)Node
```

The following figure shows the linking up of two `Leaf` trees, each having only one data element. The result is a `Node` tree:

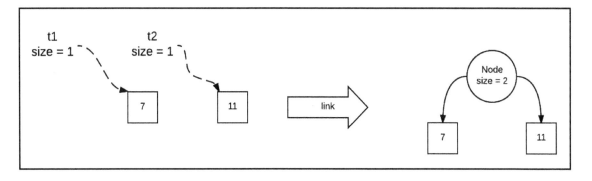

Now let's see how we could absorb a new element into a list of trees:

```scala
scala> def addTreeToList(t1: Tree, listOfTrees: List[Tree]): List[Tree] =
listOfTrees match {
     |    case Nil => List(One(t1))
     |    case Zero :: ts => One(t1) :: ts
     |    case One(t2) :: ts => Zero :: addTreeToList(link(t1, t2), ts)
     |    case _ => sys.error("how did it reach here?")
     | }
```

The first argument, namely `t1`, is a new tree. Its concrete type could be either `Leaf` or `Node`. The second argument, namely `listOfTrees`, is a list of trees, as the name indicates. We absorb `t1` into `listOfTrees`:

```scala
case Nil => List(One(t1))
```

This clause is hit when `listOfTrees` is empty, like at the very beginning of the insertion. Note its similarity with the `increment` method's first clause:

```scala
case Nil => List(1)
```

In this case, instead of `List(1)`, we generate `List(One(t1))`:

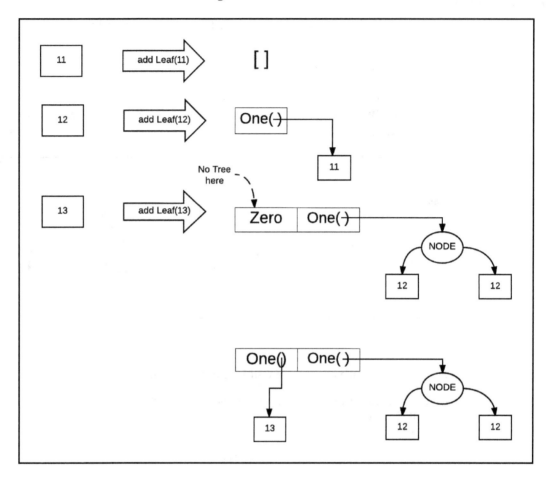

The second clause `case Zero :: ts => One(t1) :: ts` is hit when we encounter `Zero`. This is an empty slot so `Zero` is replaced with a `One` holding the `Leaf` tree. For example, in the preceding diagram, when we add the element **13**, this clause is hit.

Note the similarity with the second clause of the preceding `increment` method:

```
case 0 :: xs => 1 :: xs
```

Just like we replace the LSB 0 with 1 and structurally share the rest of the numbers list, we replace `Zero` with `One(Leaf(13))` and structurally share the rest of the tree list.

The third clause is similar to the third clause of the `increment` method:

```
case 1 :: xs => 0 :: increment(xs)
```

For the number case, we add `1` to the LSB `1`; we get `0` and a carry, which is adjusted by calling `increment` on the rest of the list:

```
case One(t2) :: ts => Zero :: addTreeToList(link(t1, t2), ts)
```

For the tree, when we get `One`, we generate `Zero` and link up `t1` with the tree `t2` within `One`.

Finally, here is the wrapper method, `cons`, that constructs the tree:

```
scala> def cons(treeList: List[Tree], x: Int) = addTreeToList(Leaf(x),
treeList)
cons: (treeList: List[Tree], x: Int)List[Tree]
```

Note the argument order: we have `treeList` first and the new element `x` second. We do this so we can do a `foldLeft` method on a collection and create the tree.

Let's use it:

```
scala> val tree = List(11, 22, 33, 44, 55, 66,
77).reverse.foldLeft(List[Tree]())((b, a) => cons
    | (b,a))
tree: List[Tree] = List(One(Leaf(11)), One(Node(2,Leaf(22),Leaf(33))),
One(Node(4,Node(2,Leaf(44),Leaf(55)),Node(2,Leaf(66),Leaf(77)))))
```

The `foldLeft` method works by taking an accumulator and a function with two arguments. Here is a refresher on how `foldLeft` works.:

```
scala> val list = List.range(1, 10)
list: List[Int] = List(1, 2, 3, 4, 5, 6, 7, 8, 9)

scala> list.foldLeft(0)((acc,x) => acc+x)
res8: Int = 45

scala> list.foldLeft(1)((acc,x) => acc*x)
res9: Int = 362880
```

A great resource on folding is available at
`https://oldfashionedsoftware.com/2009/07/10/scala-code-review-foldleft-and-fold`
`right/`.

The accumulator `acc` needs to be the first argument. As our list of trees is the accumulator, we make it the first argument.

The tree is built by inserting each element of the list successively, and the accumulator's final value is returned as the resulting tree.

Lookup

Having inserted some data, let's see how to look up a random index

Note that we must land on a `Leaf` node for the index lookup to be successful.

Again, the work is divided into two methods: `lookup`, which locates the right tree, and `searchTree`, which does a binary search within this tree.

Here is the method that performs a binary search on a tree:

```scala
scala> def searchTree (i: Int, tree: Tree): Int = (i, tree) match {
     |    case (0, Leaf(x)) => x
     |    case (i, Node(sz, t1, t2)) if i < (sz / 2) => searchTree (i, t1)
     |    case (i, Node(sz, t1, t2)) => searchTree (i - sz/2, t2)
     | }
```

The helper searches the `tree` structure for the element `i`. There are three cases:

```scala
case (0, Leaf(x)) => x
```

If `i` is 0 and the tree is a `Leaf`, then it means the search is successful and we return `i`. Pattern matching helps us write succinct code, and clauses match when both the conditions are met.

The second clause, namely `case (i, Node(sz, t1, t2)) if i < (sz / 2) =>`
`searchTree (i, t1)`, matches if `i` is placed on the left. It continues the search in the left subtree:

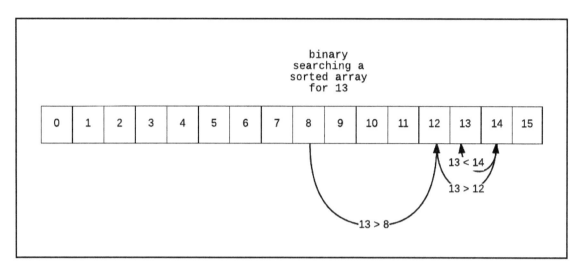

The process is very similar to how we would perform a binary search on a sorted array. The preceding figure depicts the process. Just compare it with the tree search we have.

The third clause, namely `case (i, Node(sz, t1, t2)) => searchTree(i - sz/2,`
`t2)`, matches when `i` is placed on the right subtree. The search, accordingly, continues there.

Here is the `lookup` method; it locates the right tree within the list so we can perform the binary search:

```
scala> def lookup(i: Int, tree: List[Tree]): Int = tree match {
     |    case Zero :: ts => lookup(i, ts)
     |    case One(t) :: ts if i < t.size => searchTree (i, t)
     |    case One(t) :: ts if i >= t.size => lookup(i - t.size, ts)
     | }
```

Here is a diagram illustrating the scenario:

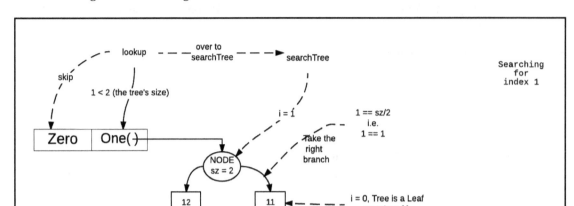

The preceding diagram depicts the various cases that crop up as the algorithm is run. Here is a case-by-case analysis.

We know that b is an empty slot. The first clause case Zero :: ts => lookup(i, ts) checks for it and just skips it, continuing with the rest of the tree list.

The second clause case One(t) :: ts if i < t.size => searchTree (i, t) is hit when i is less than the size of the tree. The index is there within the tree, so we just go and search for it via a call to searchTree(i, t).

The third clause case One(t) :: ts if i >= t.size => lookup(i - t.size, ts) skips the current One; however, it needs to adjust the index. The new index is obtained by subtracting the tree size from i.

Tracing out this scenario is left as an exercise for you.

Now let's play with the lookup operation. Note the list reversal. Can you guess why it is needed? Writing a wrapper method to hide the reverse is left as an exercise for you:

```scala
scala> val tree = List(11, 22, 33, 44, 55, 66,
77).reverse.foldLeft(List[Tree]())((b, a) => cons
     | (b,a))
tree: List[Tree] = List(One(Leaf(11)), One(Node(2,Leaf(22),Leaf(33))),
One(Node(4,Node(2,Leaf(44),Leaf(55)),Node(2,Leaf(66),Leaf(77)))))

scala> lookup(3, tree)
res0: Int = 44
```

```
scala> lookup(4, tree)
res1: Int = 55

scala> lookup(0, tree)
res2: Int = 11
```

The lookup works as expected.

Removal, head, and tail

The removal method removes a given tree from the tree list. This method works as a helper for the `head` and `tail` methods, as you will see in a minute.

By the way, the removal closely corresponds to the binary number's `decrement` method:

```
scala> def decrement(numList: List[Int]): List[Int] = numList match {
     |     case 1 :: Nil => Nil
     |     case 1 :: xs => 0 :: xs
     |     case 0 :: xs => 1 :: decrement(xs)
     |     case _ => sys.error("Not a binary number")
     | }
```

Here is a quick REPL session to use it:

```
scala> decrement(List(0, 0, 1))
res13: List[Int] = List(1, 1)
```

We decrement the binary 100 (decimal 4). Binary digits are written in a reversed fashion, so the LSB would come first.

This is similar to the `increment` method. Tracing out the execution of `decrement` is left as an exercise for you.

The `removeTree` method is based on the preceding `decrement` method. It removes the *first One(tree).*

Here it comes:

```scala
scala> def removeTree(tree: List[Tree]): (Tree, List[Tree]) = tree match {
     |    case One(t) :: Nil => (t, Nil)
     |    case One(t) :: ts => (t, Zero :: ts)
     |    case (Zero :: ts) => {
     |      val (Node(_, t1, t2), tss) = removeTree(ts)
     |      (t1, One(t2) :: tss)
     |    }
     |    case _ => sys.error("how did it reach here?")
     | }
```

Here is a diagram depicting the scenario:

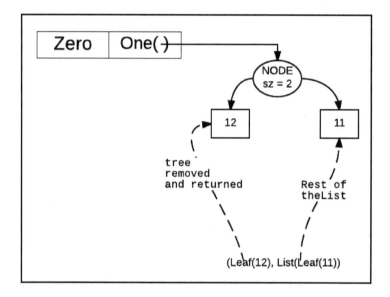

The method returns a pair: the first element is the tree being removed and the second is the list that remains after the removal.

The first case is when we have just one tree in the list:

```scala
case One(t) :: Nil => (t, Nil)
```

In this case, we return the tree and an empty list. See the `decrement` method's first clause:

```
case 1 :: Nil => Nil
```

The second clause `case One(t) :: ts => (t, Zero :: ts)` is hit when we have the first tree and the rest of the list, which is non-empty. We return the tree and the remaining list. Note the similarity with the `decrement` method's second clause:

```
case 1 :: xs => 0 :: xs
```

The third clause is this:

```
|   case (Zero :: ts) => {
|     val (Node(_, t1, t2), tss) = removeTree(ts)
|     (t1, One(t2) :: tss)
|   }
```

It is hit when we get a `Zero` element. We skip `Zero` and make a recursive call to `removeTree` with the remaining list as an argument.

We must get a `Node` in return! (Why?) The node's left tree is removed and returned as the first element of the pair. The second element is formed by prepending `t2` to the rest of the list.

Note the similarity with the third clause of the `decrement` method. We continue skipping `0` until we hit `1`:

```
case 0 :: xs => 1 :: decrement(xs)
```

Note how we can assign names to a tuple's element:

```
scala> val (one, _, three) = (1,2,3)
one: Int = 1
three: Int = 3
scala> one
res14: Int = 1
scala> three
res15: Int = 3
```

We assign a tuple with three elements to a names list, prefixed with `val`. This assigns the value 1 to the variable `one` and 3 to the variable `three`. The second tuple value is discarded by assigning it to `_`. See
`http://alvinalexander.com/scala/scala-tuple-examples-syntax` for more information.

Once we have `removeTree` defined, the `head` method is simple:

```
scala> def head(treeList: List[Tree]) = removeTree(treeList) match {
     |    case (Leaf(x), _) => x
     | }
```

As `remoteTree` always returns a `Leaf` node (why?), we just extract the `Leaf` node's data element as x and return it:

```
scala> def tail(treeList: List[Tree]) = removeTree(treeList) match {
     |    case (_, ts) => ts
     | }
```

The `tail` method is similar: it ignores `Leaf` and instead returns the rest of the list.

Let's give these methods a spin:

```
scala> val treeList = List(11, 12, 13).reverse.foldLeft(List[Tree]())((b,
a) => cons
     | (b,a))
treeList: List[Tree] = List(One(Leaf(11)), One(Node(2,Leaf(12),Leaf(13))))

scala> head(treeList)
res10: Int = 11

scala> tail(treeList)
res11: List[Tree] = List(Zero, One(Node(2,Leaf(12),Leaf(13))))

scala> head(tail(treeList))
res12: Int = 12
```

Update

How do we update an index's value? Note that we are in the *immutable, multiversioned* land.

The answer is this: we return a new version of the data structure with most of it *structurally shared* and copied just enough to reflect the change.

Its algorithm is very similar to that of the lookup operation. We locate the right tree via the `setVal` method and change the actual value via `setValInTree`:

```
scala> def setValInTree(i: Int, newval: Int, tree: Tree): Tree = (i, tree)
match {
     |    case (0, Leaf(x)) => Leaf(newval)
     |    case (_, Node(sz, t1, t2)) if (i < sz/2) => Node(sz,
setValInTree(i, newval, t1), t2)
```

```
        |   case (_, Node(sz, t1, t2)) => Node(sz, t1, setValInTree(i - sz/2,
    newval, t2))
        |  }
    setValInTree: (i: Int, newval: Int, tree: Tree)Tree

    scala> def setVal(i: Int, newval: Int, treeList: List[Tree]): List[Tree] =
    treeList match {
        |   case Zero :: ts => Zero :: setVal(i, newval, ts)
        |   case One(t) :: ts if i < t.size => One(setValInTree(i, newval, t))
    :: ts
        |   case One(t) :: ts => One(t) :: setVal(i - t.size, newval, ts)
        |  }
    setVal: (i: Int, newval: Int, treeList: List[Tree])List[Tree]
```

We will just look at the one important clause in `setValInTree`. The rest is exactly the same as `lookup`:

```
    case (0, Leaf(x)) => Leaf(newval)
```

The clause is hit when we find the index element. In the `lookup` method, we return the index value x. Here, we *forget* the value and return a *new version* of the tree where `Leaf(x)` is replaced by `Leaf(newval)`.

What could we say about the complexity of the various operations? For the operations on a binary tree, the worst case time complexity is $O(log(n))$.

What about finding the right binary tree? The key to understanding this is the binary number representation. The worst case time complexity is again $O(log(n))$.

Summary

Lists are good for operations at the head. However, for randomly accessing an element by its index, the complexity is $O(n)$.

We looked at an alternative list implementation, a list allowing fast random element access. The data structure was really a *list of binary trees*. The implementation mirrors binary number algorithms.

We saw how all the operations, such as insertion, update, lookup, head, and tail, perform at $O(log(n))$. We first locate the right tree quickly and then perform a binary search on it.

All these operations are immutable of course. We saw how the process of structural sharing and copying just enough help with the implementation.

Hopefully, this has made you hungrier for more functional data structures and algorithms. So let's look at some more exciting algorithms. Stay tuned!

8
Queues

A queue is another fundamental data structure; it is heavily used in many contexts. Queuing up tasks is a very common requirement. For example, if we look at the producer/consumer pattern, we see that both the producers and consumers communicate using a queue.

You will find many examples of queues being used in practice. Here's one of them: In an actor framework, an actor sends a message to another actor. These messages are enqueued in the recipient actor's mailbox, which is a queue.

Message brokers use queues to order and distribute messages to consumers. Priority queues are used by operating system schedulers so high priority tasks are handled before tasks that are low on priority.

In this chapter:

- We will look at functional, persistent FIFO queues. The persistent FIFO queue has an interesting design, using list prepending techniques.
- We will also look at priority queues, which are implemented using heaps. We will familiarize ourselves with the core data structure and how it is implemented in the imperative world.
- We will soon see why we need another data structure to realize functional heaps and introduce leftist trees. We will develop a functional priority queue using leftist trees.

Understanding FIFO queues

FIFO queues are used for implementing the first come, first served strategy. For example, let's consider a queue for booking movie tickets (for now we'll pretend that there is no such thing as an online ticket booking system).

The person who comes first and joins the queue gets serviced before the person after him (who joined the queue a bit later). This enforces an order, which is also known as first come, first served or FIFO. According to this, the person who comes earlier is serviced first and moves out of the queue.

A new element queues up at the head and is removed/serviced at the tail, as shown in the following figure:

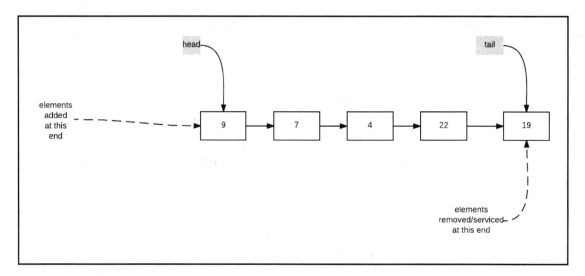

In the imperative world, it is pretty easy to see how we would implement a FIFO queue. We could maintain the FIFO queue as a singly linked list with two pointers: one at the head and another at the tail.

We will remove the tail node (pop the element off the queue)–an $O(n)$ operation–as we need to traverse the entire list to reach the last element. We will insert a new element at the head node (push the element to the queue), which would be an $O(1)$ operation. These would be simple list operations, mutating the list, to reflect the changes.

To improve the performance of the pop operation, we could instead go for a doubly linked list. This would help us pop the last element easily, thereby make it an $O(1)$ operation.

In this case, note the space (every node needs to maintain a pointer to the previous node) trade-off so we get a faster removal operation.

Functional FIFO queues

We know by now that all this mutating won't work when we deal with persistent data structures (also known as *versioned data structures*). How can we implement these queues so that when an element is enqueued or dequeued, the earlier version of the data structure would be preserved?

The design is beautiful; it involves two lists. The following diagram shows two lists, namely in and out:

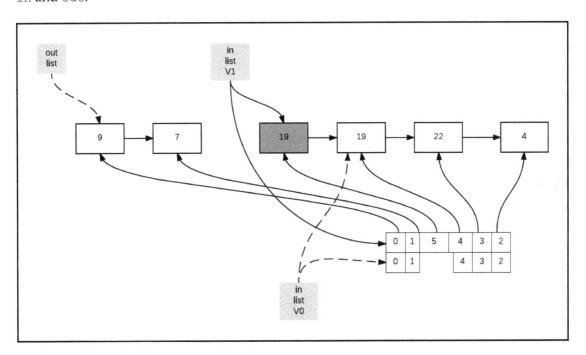

The out list holds the elements that will be popped out. We just remove the head element and return it. The in list is where new elements are inserted, that is, prepended. As we have already seen, both list prepend and head removal are *O(1)*. For example, given the preceding diagram, when we remove the element **9**, we get another version of the persistent queue: **V1**.

The following figure shows the persistence in action; note the indices for both the versions:

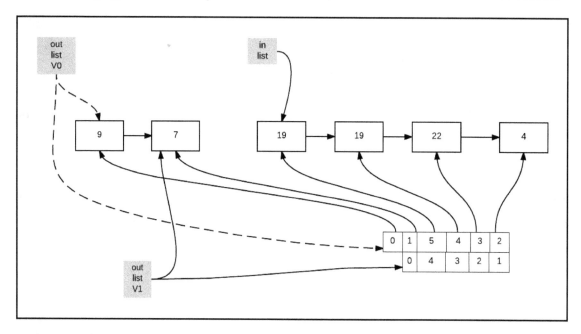

Now let's pop out another element: **7**. This will make the `out` list empty. Of course, this will create another version of the `out` list: **V2**. This will be empty, as shown in the following figure:

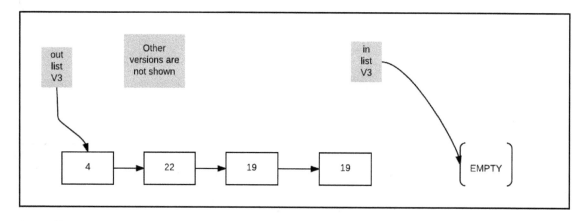

Now we have a situation where the in list is *non-empty* and the out list is *empty*. In this case, we first reverse the in list and then swap it with the out list. This is shown in the preceding diagram. This operation is called a pivot step and its cost is *O(n)*.

There is another way to look at it. We can either have both the lists empty (meaning the queue is empty) or just the in list.

We will never come across a situation where the out list will be empty and in list non-empty. This is an invariant, that is, a condition that always holds. The condition could be violated for a moment, leaving the invariant broken. In such cases, we do the needful (swap in.reverse() with out) and restore the invariant.

Invariants

Let's discuss the concept of an invariant. What is an invariant? It is a condition that always holds. For example, refer to the following code:

```
scala> val arr = Array.fill[Int](5)(-1)
arr: Array[Int] = Array(-1, -1, -1, -1, -1)

scala> for(i <- 0 to 4) {
     |    arr(i) = 0
     | }
```

To comprehend the preceding code, the following statements are always true:

- At the start of each iteration, elements at indices 0 to i-1 are zero
- After each iteration of the loop, elements 0 to i are zero

This is called a loop invariant. We can reason the code using invariants.

Another example of an invariant for a singly linked list is that every list node gets pointed at by one, and only one pointer. In other words, two adjacent pointers will never point to the same node.

Invariants can be relaxed temporarily and then restored. For example, consider the imperative version of the list node insertion algorithm:

```
q <- new node
p <- node after which q will be inserted
q.next <- p.next   // invariant temporarily violated
p.next <- q
```

Note the third step: both the adjacent pointers p and q are pointing to the same node. However, in the last step, things are put right and the invariant is restored.

Invariants translate into programming assertions, as we will soon see. Asserting the invariants will make sure our algorithms are robust.

Implementing a priority queue

For our queue implementation, the following invariant holds. If the out list is empty, the in list has to be empty (that is, the entire queue is empty). During the pivot step, the invariant is temporarily violated and then restored:

```
scala> case class Fifo(out: List[Int], in: List[Int]) {
     |
     |   def check(): Boolean = (out, in) match {
     |     case (Nil, x :: xs) => false
     |     case _ => true
     |   }
     |   require(check, "Invariant Failed - out.em")
     | }
defined class Fifo
```

Note the use of the require method to make sure the invariant always holds. We use a method called check; this method, in turn, uses pattern matching to check both the lists.

Here's the first clause:

```
case (Nil, x :: xs) => false
```

This clause matches when the out list is empty and the in list is non-empty, that is, it has at least one element. We know this violates our invariant and the object construction is aborted as a result.

This makes it impossible to end up in a state with a queue that has elements inserted but no ability to pop them out. This is an anomaly that the assertion helps us catch.

Here's the second clause:

```
case _ => true
```

It matches all other cases. This catchall matches if both the lists are empty, both are non-empty, or only the in list is empty. All these states are correct so we return `true` and the `Fifo` instance is created.

Here is an REPL session to make use of the `Fifo` instance creation:

```
scala> Fifo(List(1), List(1,2))
res0: Fifo = Fifo(List(1),List(1, 2))

scala> Fifo(List(), List(1,2))
java.lang.IllegalArgumentException: requirement failed: Invariant Failed -
out.em
  at scala.Predef$.require(Predef.scala:219)
  ... 34 elided
```

Note that the queue is represented as a `case` class. A `case` class suits us very well as it is immutable:

```
scala> val p = Fifo(List(1), List(1,2))
p: Fifo = Fifo(List(1),List(1, 2))

scala> p.in = List(1,2,3)
<console>:13: error: reassignment to val
       p.in = List(1,2,3)
```

This ensures that our contract of immutability is being adhered to automatically.

Let's look at the `push` method now:

```
scala> def push(e: Int, queue: Fifo): Fifo = {
     |    val newIn = e :: queue.in
     |    queue.out match {
     |      case Nil => Fifo(newIn.reverse, Nil)
     |      case _ => queue.copy(in = newIn)
     |    }
     | }
push: (e: Int, queue: Fifo)Fifo

scala> val q = push(1, p)
q: Fifo = Fifo(List(1),List())
```

Note that when we push the value 1 to an empty queue, it immediately ends up in the `out` list, ready to be popped out.

Here comes the `pop`:

```scala
scala> def pop(queue: Fifo): (Int, Fifo) = {
     |    queue.out match {
     |       case Nil => throw new IllegalArgumentException("Empty queue");
     |       case x :: Nil => (x, queue.copy(out = queue.in.reverse, Nil))
     |       case y :: ys => (y, queue.copy(out = ys))
     |    }
     | }
pop: (queue: Fifo)(Int, Fifo)
```

It returns a pair as a result of this: the popped value and the new, changed queue. We use the case class copy method to reflect the state changes.

As noted earlier in `Chapter 2`, *Building Blocks*, we change things only at construction time, in this case, reversing the `in` list and substituting it as the `out` list. This makes sure that if the queue is not empty, the elements will be ready to get popped out:

```scala
scala> val q = push(3, push(2, push(1, Fifo(Nil, Nil))))
q: Fifo = Fifo(List(1),List(3, 2))

scala> pop(q)
res6: (Int, Fifo) = (1,Fifo(List(2, 3),List()))
```

Note that in this case, the act of popping made the `out` list empty.

Understanding priority queues/heaps

Priority queues are queues where each element has a priority. An element with high priority is served before an element with low priority.

For example, consider we have a task queue where tasks are inserted and need to be executed. A high priority task may appear after some tasks are inserted in the queue; however, it would need to be executed prior to tasks with low priority.

There are min-heaps and max-heaps. Min-heaps always have the least element as their root, which would be readily accessible. For max-heaps, the max element will be the root.

Let's look at the min-heap data structure first and then the functional version. Heaps are complete binary trees.

For more on the definition, visit
`http://web.cecs.pdx.edu/~sheard/course/Cs163/Doc/FullvsComplete.html.`

The nodes also have partial ordering such that the root value is always less that its children. You will get this if you've been a keen observer all this while: this is an invariant.

In the imperative world, heaps are represented with an array. This representation makes heaps an implicit data structure. We don't need any auxiliary space, such as a pointer, to store the parent-child relationship. As shown in the following figure, the left child of an element at index i is stored at $2*i + 1$. The right child is stored at $2*i + 2$.

In a Binary Search Tree, the left child's value is always less than that of the right child (and the root). In this case, however, there is no specific order in the left and right children. Let's have a look at the figure:

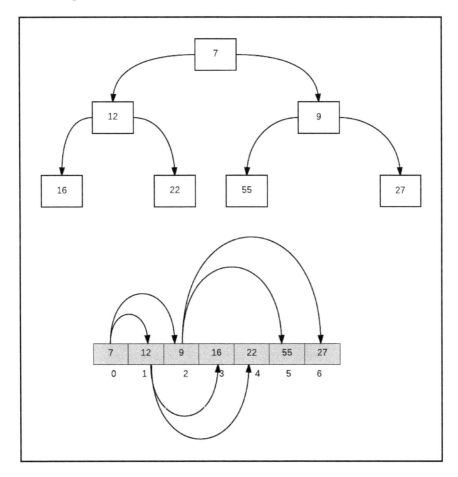

The `min_elem` operation has *O(1)* complexity, as the min element is readily available at the root

What about inserting and popping elements? The root value is removed and is overwritten by the rightmost element. This could momentarily violate the invariant, so some swapping is required to restore it.

Let's have a look at the figure:

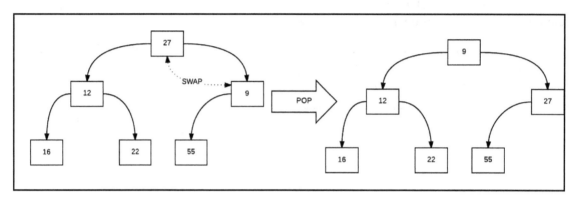

As shown, we overwrite the value at root **7** with **27**, which is the bottom rightmost element.

Now this operation leaves the tree in an inconsistent state, as the invariant no longer holds, **27** being higher than both **12** and **9**.

So to restore it, we look at both the children and swap them with the smaller value. So **27** and **9** are swapped and we check again for the subtree at **27**. As **27** < **55**, the invariant holds and we have the *heap property* restored. Note that if the value were instead lower than **27**, one more exchange would be required.

This process is very similar to the search operation in BSTs. The complexity of the pop operation is thus *O(n)*.

Insertion is very similar–we add elements to the end of the array instead.

For more information, visit
`https://www.cs.cmu.edu/~adamchik/15-121/lectures/Binary%20Heaps/`
`heaps.html`.

Leftist trees

Think about the problem we'd face if we try to adapt this array-based algorithm to a persistent version. The swap will result in expensive copying, so an insert/pop would have complexity amounting to $O(n)$.

A leftist tree is a data structure that we can use to implement the priority queue ADT. Before you look at the core data structure, look at the *rank* of the tree.

We first make the tree a full binary tree. If we put explicit leaves in such a tree, every node (other than the leaves) will have two children.

For more information, visit:

```
http://stackoverflow.com/questions/12359660/difference-between-complete-binary-
tree-strict-binary-tree-full-binary-tre
```

Let's have a look at the figure now:

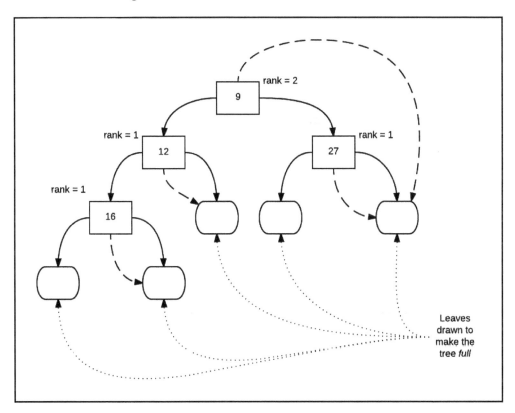

We define the *rank* of a binary tree as per the length of its *right spine*. The rank of the leaf is **0**. In the preceding figure, the rank of the tree at root **9** is **2** as we need to cross over the right node with value **27**. The right spine for every node is drawn with a dashed line.

A leftist tree is one that satisfies an invariant. For every subtree t, the following property must hold:

rank(left(t)) >= rank(right(t))

This invariant also goes by the name the leftist condition.

Let's check the preceding tree. The rank of the left subtree at **12** is equal to the rank of the sibling right subtree at **27**. The rank at node **16** is **1**, which is higher than the sibling (which is an empty tree) with rank **0**.

So we conclude that the preceding tree satisfies the invariant and is a leftist tree. On the other hand, the following is not a leftist tree:

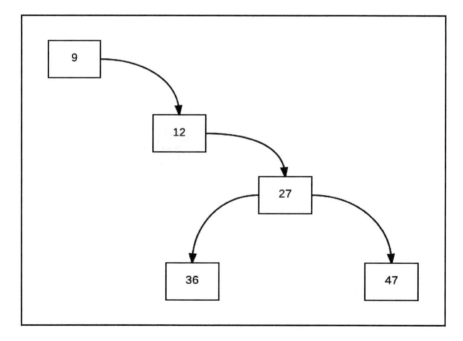

We need to do the following transformations to make it leftist. Note that the subtree rooted at **27** is leftist; however, both the tree and subtree rooted at **12** and **9** (in the preceding figure) are not. Let's have a look at the figure:

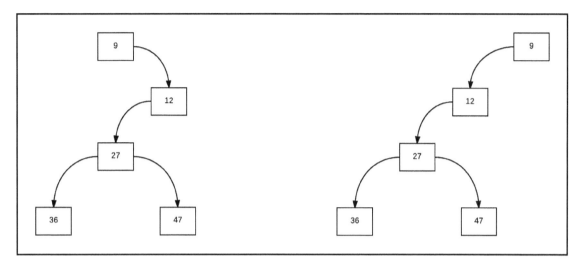

The diagram shows the following two transformations:

1. Swap the left and right children of **12**.
2. Next, swap the left and right children of **9**.

This restores the leftist invariant and the tree becomes leftist.

Functional heaps

Note that if we observe the heap invariant–that is, root elements are never greater than child values–on a leftist tree, we will get a leftist heap.

The preceding tree, for example, is a leftist heap. A notable property of a leftist heap is any path is a sorted list. This helps us efficiently merge the tree after a pop operation.

Where does merge come from? We know the minimum element is at the root. So, when we pop or remove the root element, we will get two leftist trees. If we merge these two, we get back to a sane state (invariants restored) and get the next version of the persistent heap.

The case for inserting a new node could be expressed as a merge again. The new node can be looked at as a singleton tree (a tree with just one node). This is merged with the existing tree, which adds the new node as a result of this. Here comes the code:

```
sealed abstract class TreeNode {
  def rank: Int
}
```

The `sealed` keyword makes sure we know all the subclasses, as these need to be in the same source file. For example, `Option` is a sealed class.

Sealed classes help make sure the pattern matching is exhaustive. We use pattern matching a lot, and this is a very helpful aid. For example, have a look at the following code:

```
scala> def matchIt(y: Option[String]) = y match {
     |    case Some(k) => println(k)
     | }
<console>:10: warning: match may not be exhaustive.
It would fail on the following input: None
         def matchIt(y: Option[String]) = y match {
                                            ^
```

This is a very good flag, warning us in advance whether we have missed some leg of the pattern match:

```
case class Node(rank: Int, v: Int, left: TreeNode, right: TreeNode) extends
TreeNode
```

Checking for the leftist invariant is surely a good idea and is left as an exercise for you.

The `Node` class is a kind of `TreeNode`, holding a rank, a value, and the left and right children:

```
case object Leaf extends TreeNode {
  override val rank = 0
}
```

This is our `Leaf` class, which is a `case` object. (In earlier chapters, we used the `case` object to implement the `List` node's `Nil` and the `BinTree` node's `Leaf` nodes.) A `Leaf` instance does not have any state of its own; instead, it works as a kind of terminator for our tree.

This is clearly a case for a singleton. We need just one instance of the `Leaf` class. As an aside, we should always use the `case` object when our `case` class constructor has no arguments. You just don't need multiple instances of an immutable and stateless class. One would suffice; this avoids needless creation of other instances:

```
def makeNode(v: Int, left: TreeNode, right: TreeNode): Node =
  if (left.rank >= right.rank) Node(left.rank + 1, v, left, right)
  else Node(right.rank + 1, v, right, left)
```

When a new node for our persistent tree is manufactured, we make sure the leftist tree invariant is observed. The rank of the left child is higher than or equal to the right child's rank. If it is not so, we switch the left and right children so the invariant is maintained:

```
def merge(node1: TreeNode, node2: TreeNode): TreeNode = (node1, node2)
match {
  case (node1, Leaf) => node1
  case (Leaf, node2) => node2
  case (Node(_, x, a1, b1), Node(_, y, a2, b2)) => {
    if (x < y) makeNode(x, a1, merge(b1, node2))
    else makeNode(y, a2, merge(node1, b2))
  }
}
```

This is a workhorse method that merges with leftist trees. Refer to the following diagram that illustrates how `merge` operates. We have two leftist trees, denoted by **x** and **y**. For illustrations sake, we show a flow using an explicit stack. The code has an implies stack due to recursion.

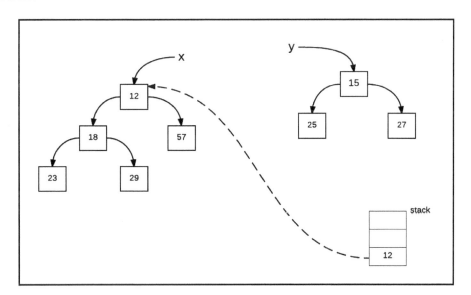

We remember the subtree with the lesser value; we do this by pushing it onto the stack and advancing to the right of the least value.

As **x** tree's value was pushed, we advance **x** to the right node, with value **57**, compare again and remember (that is, `push`) the subtree with the smaller value.

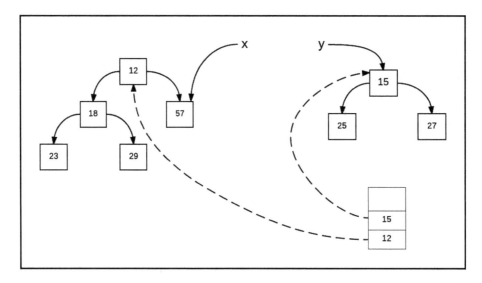

We now move **y** (as it was smaller and pushed to stack) to its right child with value **27**:

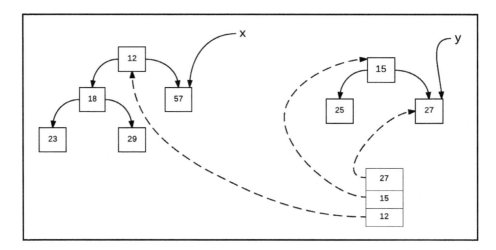

If we repeat again, **y** tree's right child is a leaf. In this case, we take the other argument: pointing at **57** and pushing it. Once we hit a leaf, we pop the remembered subtrees and link them up.

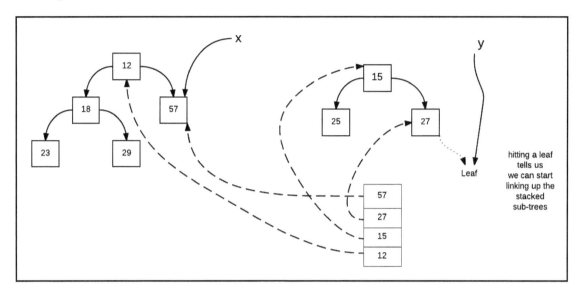

How do we link the remembered subtrees? We do the following algorithm:

```
m <- pop the stack
make sure m is leftist (swap left and right if required)
if (stack is not empty) {
    t <- top of stack // do not pop
    t.right <- m
} else {
    return m
}
```

The following figure illustrates the case when the subtree at **57** becomes the right subtree of **27**:

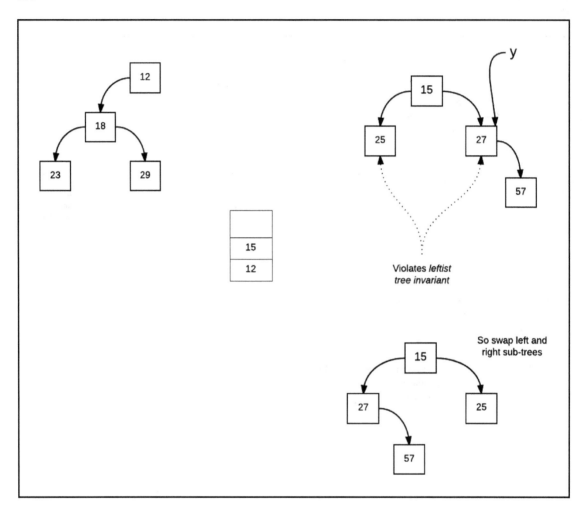

We continue with the stack, remove **15**, and make it a right child of **12**:

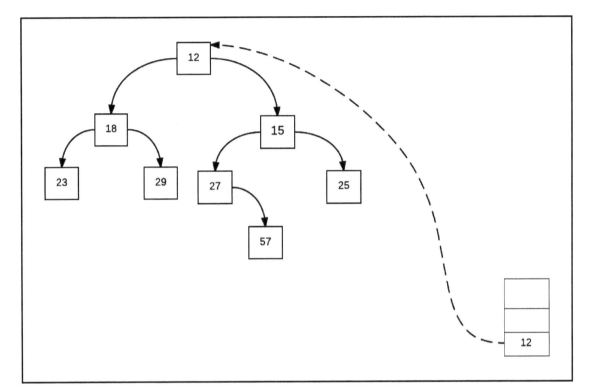

We check whether the tree at **12** is a leftist. It is, so we return **12** as the new root and the merge is completed.

The rest of the methods are simple once you understand how `merge` works:

```
def insert(v: Int, h: TreeNode): TreeNode = {
  val singleTree = makeNode(v, Leaf, Leaf)
  merge(singleTree, h)
}
```

The insert method takes a value v and a leftist tree h. It turns the value v into a leftist tree with just one node, a singleton tree.

It calls the `merge` function on both the trees and returns a new version of the tree with the value v added:

```
def min (h:TreeNode): Int = h match {
  case Leaf => throw new IllegalArgumentException ("Empty tree")
  case (Node(_, x, _, _)) => x
}
```

The `min` method is pretty simple. We pattern match the argument to check whether it is `Leaf` or `Node`. A `Leaf` signifies an empty tree, so there is no min that exists and hence we throw an exception.

A `Node`, on the other hand, always has a value. We use Scala's mighty underscore to ignore the rank, left and right fields, and just pick up the value and return it:

```
def pop(h:TreeNode): (Int, TreeNode) = h match {
  case Leaf => throw new IllegalArgumentException ("Empty tree")
  case (Node(_, x, a, b)) => (x, merge(a, b))
}
```

The `pop` method pops the min element of the heap. It then merges the left and right subtrees, which are leftist trees themselves, thereby forming a newer version of the leftist tree with the earlier root node removed.

Summary

Like stacks, queues are fundamental data structures. Queues help us realize the FIFO approach. We looked at how FIFO queues are implemented in the imperative world. We noted that we would end up with too much of copying if we insert new nodes at the end of a list. We could do this in the imperative world but would end up with $O(n)$ copying performance to achieve persistent FIFO queues.

Instead, we looked at an innovative design involving two stacks. We also looked at the Scala implementation and discussed some Scala idioms.

Priority queues are an important variation of queues. We define a *priority* for each element of the queue and wish to pop the element with the highest priority.

Heap is a famous data structure for implementing the priority queue ADT. Heaps are realized with a full and complete binary tree. This is not a BST though. The heap invariant is this: the value at the root is less than its children.

We looked a beautiful algorithm, based on arrays, to implement heaps. However, this again does not work in the functional setting as array swaps are costly.

An alternate data structure, a leftist tree, is also used to implement heaps. The invariant is that the rank of the left subtree is always equal or higher than the right subtree. This data structure has nice performance characteristics in a functional setting.

We accordingly developed the functional heaps using leftist tree. We did a Scala implementation and looked at the related Scala idioms.

With the stage thus set, we will look at *double ended queues,* also known as, *dequeues* in the next chapter. In a dequeu, insertion and deletion can happen at both ends.

Welcome to yet another fun ride!

9
Streams, Laziness, and Algorithms

You have visited a new city to attend an important business meeting. You started your journey early in the morning because this meeting is very important for business and you want to be on time for meeting. You reached the new town and started to feel hungry. So, you looked for some restaurants and, fortunately, you found one with a very good ambiance and soothing instrumental music. As you started flipping the menu, you were attended by a waiter. Very soon you found your favorite dish. The waiter took the order and, to your surprise, the food had been served within few seconds. Somehow, you tasted the food and found that it was tasteless and not been prepared fresh. You complained about it and asked the waiter for fresh food. But the waiter replied that they always cook the food in the morning to be served later, the whole day. Somehow, you ate the food and moved on for the meeting.

In the meeting, you were asked to present sales reports of the company product in your area. At that time you are finding it difficult to recall the actual data and calculations, and realized to redo all the hard work again, which made you more tired and frustrated. Finally, you did the recalculation and provided the result.

In the evening, after returning home, all these incidents made you to realize the power of *on-demand* preparation, that is, if food might have been prepared on demand, you might have tasted fresh food and the restaurant might have utilized its resources optimally.

The second thing which bothered you was that forgetting the lengthy calculation could be avoided if you had recalled the result of the previous calculation.

In this chapter, we are going to learn lazy evaluation (on-demand computation), memoization (remembering past results), and stream (an infinite list).

Program evaluation

We write computer program in one or more files. These files are known as source code. When source code is executed and run in memory, it is known as process. Process executes each program statement one by one. Execution of statements is known as evaluation. There are two types of evaluation strategy, *eager evaluation* and *lazy evaluation*. Later in this chapter, we will explore and understand types of evaluations. Let's start with eager evaluation.

Eager evaluation

An expression is executed whenever it comes on the way of program execution. This evaluation strategy is known as **eager evaluation**, which is also known as *strict evaluation*. Even the value of expression is not required at the moment the expression is evaluated.

Let us consider the following code:

```
scala> var fv  = 5

fv: Int = 5
```

We have typed `var fv = 5` in console. As it is executed, value 5 is created and bound to variable `fv`.

But, this is not the only case. In some conditions, expressions or part of expressions are only evaluated when they are needed.

We can understand on-demand computation using AND operator. The AND operator employs short-circuit evaluation. Following is the case of minimal evaluation:

```
def leftBool(x : Int) : Boolean ={
    println("Left Function")
    x > 5
    }
```

The `leftBool` function will take an integer as an argument and return `true` if the supplied integer is greater than 5. Before returning `true` or `false`, it will print the message `Left Function`:

```
def rightBool(x : Int) : Boolean = {
    println(" Right Function")
    x < 6
  }
```

The `rightBool` function has only one integer argument, and it returns `true` if the integer is less than 6. Similar to the `leftBool` function, the `rightBool` function, before returning the `true` or `false`, will print the message `Right Function`:

```
scala> val x = 3

x: Int = 3

scala> leftBool(x) && rightBool(x)

Left Function
res0: Boolean = false
```

Variable `x` is associated with a value of 3 then AND operator is called with the `leftBool` and `rightBool` functions. The `leftBool` function return `false`; hence, the AND operator does not need to evaluate the `rightBool` function therefore, `rightBool` function is not evaluated:

```
scala> val x = 6

x: Int = 6

scala> leftBool(x) && rightBool(x)

Left Function
Right Function
res1: Boolean = false
```

Now x is associated with value 6. The `leftBool` function returns `true`; hence, `rightBool` is evaluated.

Here, the `rightBool` function is only evaluated when it is needed to provide the value.

Discussing and exploring on lazy evaluation is going to be very interesting. In the following section, we will do a detailed study of lazy evaluation and its characteristics.

Argument evaluation

When we call a method (or a function), we could ask for *delayed argument evaluation*. Following is an example Scala snippet:

```
scala> def aMethod(x: String) =
     |    println(x)
aMethod: (x: String)Unit

scala> aMethod("Hello world") // call site
Hello world
```

The method argument is *evaluated* at the call site. However, recall that functions are *first class values*. This allows us to write the method as follows:

```
scala> def aMethod_1(x: () => String) =
     |    println(x()) // argument evaluated
aMethod_1: (x: () => String)Unit

scala> aMethod_1(() => "Hello world") // unevaluated argument
Hello world
```

When `aMethod_1` is called, the string argument is not evaluated at the call site. Inside the method definition, the function is called and the argument value is obtained and printed. We are delaying the evaluation till the value is really needed!

An argument evaluation is delayed just by putting `()` `=>` immediately before its type. We essentially convert the type from `A` to `()` `=>` `A`, that is, a function that accepts zero arguments and returns an `A`.

This function, when used to realize delayed evaluation, is called a **thunk**. In the preceding snippet, expression `()` `=>` `"Hello world"` is a thunk.

When we call the function to finally evaluate the argument, the *thunk is forced*.

We will come across the phrase *forcing evaluation*, all it means is that the function (aka thunk) is evaluated to get the argument.

The following Scala syntax makes it easier to specify delayed evaluation for arguments:

```
scala> def aMethod_2(x: => String) =
     |    println(x)
aMethod_2: (x: => String)Unit

scala> aMethod_2("Hello world")
Hello world
```

If we put a => before the type, Scala wraps it up as a thunk. Now, instead of specifying the argument as () => A, we can just specify the argument value.

Lazy evaluation

The computer program evaluation strategy, in which a program expression is executed when needed, is known as lazy evaluation. Lazy evaluation is also called as *call-by-need* and *non-strict evaluation*. The value of executed expression is cached for subsequent use. It means that if an expression is executed lazily, then the output of expression execution is put into memory and used the value in subsequent calls of that expression.

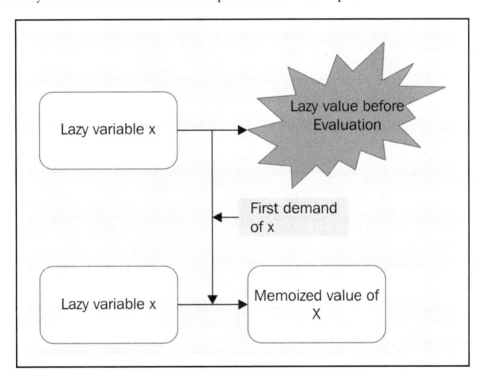

The preceding diagram is clearly depicting the nature of a lazy variable. At the time of declaration, it has not been evaluated and bounded to variable **x**. When first call is made by the program, it is evaluated and bounded to variable **x** and memoized.

Lazy evaluation in Scala

In Scala, we can initiate a lazy variable using Scala keyword `lazy`. But before we start to explore `lazy`, for newcomers to Scala, it is better to understand type of variable in Scala. Scala has two types of variable.

- Variables whose value can be changed—these variables are created using the `var` keyword:

```
scala> var sv = 5

sv: Int = 5

scala> sv = 6

sv: Int = 6
```

   The `sv` is a `var` type variable. Therefore, the value of the variable can be changed as many times as you want to change it.

- Variables whose value cannot be changed—these variables are created using Scala keyword `val`:

```
scala> val svc = 5
svc: Int = 5

scala> svc = 6

<console>:8: error: reassignment to val
      svc = 6
          ^
```

The `svc` is a `val` type variable. As a result, its value can be associated only at the time of declaration of the variable. Furthermore, its value will be considered as constant and it cannot be changed.

In the following examples, we will explore the properties of keyword `lazy`:

```
scala> val sv = 5

sv: Int = 5
```

The sv is a simple variable. When the statement val sv = 5 is executed at the same time, value 5 is associated with variable sv. Therefore, REPL is shows associated with sv on the console.

The following example will throw some light on lazy evaluation:

```
scala> lazy val svl = 5

svl: Int = <lazy>
```

It is clear that when we use keyword lazy, on the right-hand side of the output shows <lazy>. It is showing that the expression on right-hand side has not been evaluated:

```
scala> svl

res6: Int = 5
```

Call on svl is getting right-hand side evaluated. Hence, res6 contains the value 5. The value of svl is cached and used in subsequent calls.

```
scala> lazy var svl = 5
```

The output of preceding command is as follows:

```
<console>:1: error: lazy not allowed here. Only vals can be lazy
        lazy var svl = 5
             ^
```

The preceding code shows that we cannot define a lazy variable as a var variable. We can only define lazy variables as a constant variable using val.

Lazy evaluation in Scala will be very clear by following the next code example:

```
scala> import  java.util.Calendar   //Line One

import java.util.Calendar

scala> lazy val lazyDate = Calendar.getInstance.getTime   //Line Two

lazyDate: java.util.Date = <lazy>

scala> val simpleDate = Calendar.getInstance.getTime  // Line Three

simpleDate: java.util.Date = Sun Jun 19 05:12:48 IST 2016

scala> val rl = 1 to 20 //Line Four
```

```
rl: scala.collection.immutable.Range.Inclusive = Range(1, 2, 3, 4, 5, 6, 7,
8, 9, 10, 11, 12, 13, 14, 15, 16, 17, 18, 19, 20)

scala> for(x <- rl) print(x) //Line Five

1234567891011121314151617181920

scala> println("Simple Date Variable Value Is : " + simpleDate) // Line Six

Simple Date Variable Value Is : Sun Jun 19 05:12:48 IST 2016


scala> println("Lazy Date Variable Value Is : " + lazyDate) // Line Seven

Lazy Date Variable Value Is : Sun Jun 19 05:13:21 IST 2016
```

The preceding code snippet is clearly showing how lazy value is evaluated. *Line One* is just to `import java.util.Calendar` class. In *Line Two*, `Calendar.getInstance.getTime` will get the current date for lazy variable `lazyDate`. Current date is again created for simple variable `simpleDate` in *Line Three*. Remember that the date is created in simple variable after lazy variable date creation. When current date is in the simple variable, it is shown in variable at the same time. I have created *Line Four* and *Line Five* just to kill time. *Line Six* is printing simple date variable value, which is the same as it got in *Line Three*. Finally, in *Line Seven*, lazy date variable `lazyDate` is printed. For lazy variable, the expression `Calendar.getInstance.getTime` is executed when the `println` function is called on it.

Lazy evaluation in Clojure

Similar to Scala, Clojure has a `delay` macro to create a lazy variable. Macro `delay` will delay the evaluation of an expression. It evaluates an expression when it is demanded. Result of delayed expression is cached so that it can be used further without execution of the same expression. In the following code snippet, we will understand lazy evaluation in Clojure:

```
user=> (def lazyVal (delay 10))
#'user/lazyVal
```

The preceding code line is creating a lazy variable `lazyVal` and binding it with 10 lazily using `delay` macro:

```
user=> lazyVal
#object[clojure.lang.Delay 0x4a2c3477 {:status :pending, :val nil}]
```

Now `lazyVal` is has the `pending` and `nil` in it.

```
user=> (realized? lazyVal)

false
```

The `realized?` component will tell you if value of some variable has been realized or not. In our case, if it returns `false`, then it means that the value of `lazyVal` has not been realized:

```
user=> (class lazyVal)

clojure.lang.Delay
```

Type of variable `lazyVal` can be found using class in Clojure. It shows that `lazy val` is of the `clojure.lang.Delay` type:

```
user=> (deref lazyVal)

10
```

The macro `deref` can be used to force `delay` variables if not already forced:

```
user=> lazyVal

#object[clojure.lang.Delay 0x4a2c3477 {:status :ready, :val 10}]
```

We can see that just putting `lazyVal` is not going to print the value of `lazyVal`. Whenever the value is required, we have to dereference it using `deref` macro.

```
Let us inspect the result of realized? on lazyVal.

user=> (realized? lazyVal)
true
```

The realized? on dereferenced delayed variable `lazyVal`, returns `true`. And as the variable `lazyVal` has been dereferenced, we found `true` as result, when realized? is operated with `lazyVal`:

```
force and @ can be used also to dereference the delayed variable.

user=> (def lazyVal (delay 10))

#'user/lazyVal

user=> (force lazyVal)
```

```
10

user=> (def lazyVal (delay 10))

#'user/lazyVal


user=> @lazyVal

10
```

Clojure's promise can also be used to create a lazy variable. It shares many characteristic with delay. It differs from delay in that it requires to deliver before dereferencing the promise variables. In the following code example, we can see how to use promise to create lazy variables:

```
user=> (def fp (promise))

#'user/fp

user=> (realized? fp)

false

user=> (deliver fp (str "MyString"))

#object[clojure.core$promise$reify__7005 0x5e80fadc {:status :ready, :val
"MyString"}]

user=> (realized? fp)

true

user=> fp

#object[clojure.core$promise$reify__7005 0x5e80fadc {:status :ready, :val
"MyString"}]


user=> (deref fp)

"MyString"
```

The following example will add to our understanding of lazy evaluation in Clojure:

```
user=> (def lazyDate ( delay (new java.util.Date)))
#'user/lazyDate
```

A new lazy variable `lazyDate` is defined to hold a date type of variable:

```
user=> lazyDate

#object[clojure.lang.Delay 0x41f9f647 {:status :pending, :val nil}]
```

After creating a lazy variable `lazyDate`, now we are going to create a simple variable `simpleDate`, whose value will be realized at the time of its creation:

```
user=> (def simpleDate  (new java.util.Date))

#'user/simpleDate

user=> simpleDate

#inst "2016-06-19T00:08:52.248-00:00"
```

In the following line of code, I am just counting a list so that some computation time will be spent. Then we will test what value our `simpleDate` and `lazyDate` are printing:

```
user=> (count [1,2,3,4,5])

5
```

Now let us print `simpleDate` again. We find that value of `simpleDate` is the same as printed before:

```
user=> simpleDate

#inst "2016-06-19T00:08:52.248-00:00"
```

Now:

```
user=> (deref lazyDate)

#inst "2016-06-19T00:10:00.824-00:00"
```

We dereferenced `lazyDate` and get the final realized value.

We can observe that after dereferencing, `lazyDate` provides date at the time of dereferencing, whereas `simpleDate` provides date value at the time of initialization.

Memoization – remembering past results

Memoization is the art of computer program optimization to speed up functions. Donald Michie coined the word *memoization*. Whenever, a memoized function is called for the first time, for a given input, its output value is calculated and cached. Next time when the same input is given as an argument, function does not compute the value but returns the value from cached location for that given input. In some programming language, we find some or other internal mechanism to implement memoization. But many programming languages require explicit work to implement memoization.

So, is it possible that a function can return different output for the same input? The answer is: this is possible. We can understand this from the following Scala example:

```
scala> val ab = scala.collection.mutable.ArrayBuffer(1,2,3,4)   \\ Line One

ab: scala.collection.mutable.ArrayBuffer[Int] = ArrayBuffer(1, 2, 3, 4)

scala> print(ab.remove(1))   // Line Two
2
scala> print(ab.remove(1))  // Line Three
3
scala> print(ab.remove(1))  // Line Four
4
```

The preceding code is easy to understand. In *Line One*, an `ArrayBuffer` is created with two values in it, that is, `(1,2,3,4)`. The `ArrayBuffer` is a mutable sequential collection in Scala. Then, in *Line Two*, we use `remove` function, which is removing data at index one from `ArrayBuffer` `ab`. The `print` function prints the removed value, which is the one in *Line Two*. Again, in Line three, `print` is prints 3 for the same input as before. Hence, it is clear that for the same input, a function can give different output. Now we will consider a new example in the following code lines:

```
scala> class Multiply {
            def mult(num: Int) : Int = num*2
       }

defined class  Multiply

scala> val myMult = new Multiply()   \\Line One

myMult: multiPly = Multiply@6c6017b9


scala> print(myMult.mult(4))   \\Line Two
```

```
8
scala> print(myMult.mult(4))    \\Line three

8
```

We create a class `Multiply`, which has a method `mult`. The `mult` method will take any integer value, multiply that input value by 2, and return the output. It can be observed easily that the `mult` function for input 4 always returns 8. We know that the `mult` function is an example of referentially transparent function. So, from this example, it is very clear that we can cached the output of the `mult` function for some input. And, for the same input, we can use the memoized value of the function rather than calculating it. Memoization requires caching function outputs for different inputs in memory. Caching values look for some amount of memory. But, the speed of function is increased as its calculation is not required for input values used in the past. Generally, memoization is done using a lookup table, in which the function arguments and the corresponding values are saved.

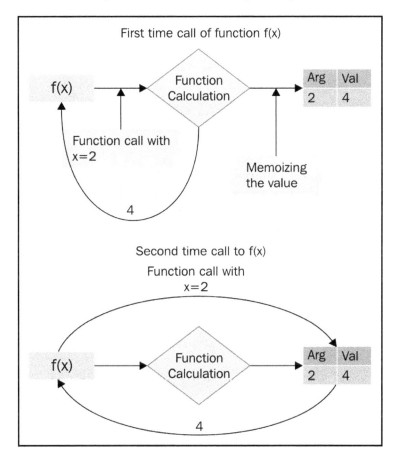

The diagram is considering a function $f(x) = x^2$. Function **f(x)** is a memoized function. When **f(x)** is being called for the first time for **x=2**, it calculates the output **4** and returns it to the caller, as well as memoizing the output of the function in a lookup table. On subsequent calls for argument **2**, the **f(x)** is not computed, but its value is taken from a lookup table and returned.

We will discuss the calculation of a factorial by a simple way and memoized way.

Memoization in Scala

Some programming languages provide built-in mechanism to memoize functions. But in Scala, we have to provide a mechanism to memoize function output values.

In the following discussion, we will explore memoization in Scala:

```scala
class SimpleFactorial{
    def simpleFactFun(num : Int) : Int ={
        if(num == 0 || num == 1)
            1
        else{
        println("Calculating Factorial")
            num*simpleFactFun(num-1)
            }
    }
}
```

The `SimpleFactorial` class is a simple Scala class. The `simpleFactFun` method will calculate the factorial of an integer recursively. Whenever the `simpleFactFun` method calculates a factorial of a number greater than 2, it will print the message `Calculating Factorial`:

```scala
scala> val simpleFact = new SimpleFactorial()

simpleFact: SimpleFactorial = SimpleFactorial@20b52576

simpleFact is an object created of class  SimpleFactorial.

scala> var factValue = simpleFact.simpleFactFun(5)

Calculating Factorial
Calculating Factorial
Calculating Factorial
Calculating Factorial
factValue: Int = 120
```

We want to calculate the factorial of an integer 5. As many times the `simpleFactFun` function has been called in recursion process, that many times message `Calculating Factorial` has been printed out on console:

```
scala> var factValue = simpleFact.simpleFactFun(5)

Calculating Factorial
Calculating Factorial
Calculating Factorial
Calculating Factorial
factValue: Int = 120
```

We again call the method `simpleFactFun` with argument 5. As a result, we observed that whenever we call the function `simpleFactFun`, it does all the computation to provide the output.

Now we will try to investigate how memoized functions can help us to write faster programs:

```
class MemoizedFactorial{
    var memoized : Map[Int,Int] = Map()

    def lookupFunction(id : Int) :Int = memoized.getOrElse(id,0)

    def memoizedFactFun(num : Int) : Int ={
        if(num == 0 || num == 1)
            1
         else if (lookupFunction(num) > 0)
             lookupFunction(num)
        else{
            val fact = num*memoizedFactFun(num-1)
        println("In memoizing part")
            memoized += num -> fact
            fact
            }
    }

}
```

The `MemoizedFactorial` class holds two methods. The `lookupFunction` will look for memoized outputs of the `memoizedFactFun` method. Outputs from `memoizedFactFun` for different integers are memoized in a `Map` type field memoized. The `memoized` variable will store output of the `memoizedFactFun` method in the form *2 -> 2, 3 -> 6*:

```
scala> var memFact =  new MemoizedFactorial()
```

We have created a new object of the `MemoizedFactorial` class:

```
scala> var factValue = memFact.memoizedFactFun(5)

In memoizing part
In memoizing part
In memoizing part
In memoizing part
factValue: Int = 120
```

We have calculated the factorial of 5 and got the result. The message `In memoizing part` is being printed as many times as the `memoizedFactFun` method is called in recursion process:

```
scala> var factValue = memFact.memoizedFactFun(5)

factValue: Int = 120
```

Recalling method `memoizedFactFun` has not printed any message. It happens because the last time when `memoizedFactFun` was called, the output value was memoized:

```
scala> var factValue = memFact.memoizedFactFun(4)

factValue: Int = 24

scala> var factValue = memFact.memoizedFactFun(3)

factValue: Int = 6
```

We can observe easily that even for smaller integers than 5, the `memoizedFactFun` method has not printed any message. It happens because while calculating the factorial of 5, the factorials of other integers less than 5 were also memoized.

Memoization in Clojure

Clojure has an internal tool to memoize a function. In Clojure, we use `memoize` to memoize a function. A referentially transparent function can be memoized using `memoize`:

```
(defn simpleFactFun [n]
  (if (or (= n 1) (= n 0)) 1
    ( do
        (println "Calculating Factorial")
          ( * n (simpleFactFun (dec n))))
  )
)
```

In order to create a memoized function, we first create a simple function `simpleFactFun` to create the factorial of an integer n. The factorial is calculated in a recursive way:

```
user=> (simpleFactFun 5)

Calculating Factorial
Calculating Factorial
Calculating Factorial
Calculating Factorial
120
```

Memoizing simpleFactFun

In the following, we are going to memoize `simpleFactFun` using the `memoize` keyword:

```
(def  memoizedFactFun (memoize simpleFactFun))
```

Using the `memoized` function of Clojure, the `simpleFactFun` function is being memoized, as shown in the preceding code line:

```
user=> (memoizedFactFun 5)

Calculating Factorial
Calculating Factorial
Calculating Factorial
Calculating Factorial
120

user=> (memoizedFactFun 5)

120
```

It can be discovered from the preceding program that the output of the `memoize` function is cached. The second time the `memoizedFactFun` function is called with the same argument, it does not print any message. This is because the memoized values are taken.

Streams

A stream is generally a lazy and linear sequence collection. Stream elements are evaluated only when needed. Streams are already defined in Scala and Clojure. Since elements of a stream are evaluated lazily, it can be of infinite length. In this section, we will explore streams in Scala and Clojure. Let me start with stream in Scala. List, which we are going to explore in the coming chapter, differs from stream only in lazy computation. Other properties of list are similar to stream.

Stream in Scala

Stream is a class defined in Scala. Stream constructor can be used to create Stream object. Stream elements are evaluated when they are required. Stream-calculated values are cached so that they can be used further. Let us explorer the characteristics of streams in Scala:

```
scala> import scala.Stream

import scala.Stream

scala> var ms = Stream("a","b","c","d","e")

ms: scala.collection.immutable.Stream[String] = Stream(a, ?)

scala> print(ms)

Stream(a, ?)
```

We have created a stream of finite length using `Stream()`. But why does `print(ms)` shows us `Stream(a, ?)`?

It is clear that the `print` function is just showing the first element, and for the rest of the elements, it is showing ?. The character ? is telling us that only the first element of the stream has been evaluated. And the remaining elements will be evaluated only when required.

Indexing the elements of a stream

Elements of a stream can be indexed using `()`. For example, if I want to get the first element of the list, the following code snippet will do it:

```
scala> ms(0)

res14: String = a
```

We have got the first element of the stream and have understood that indexing will be started from 0 not from 1 as some programming languages support:

```
scala> ms(2)

res15: String = c

scala> print(ms)

Stream(a, b, ?)
```

The `ms(2)` has printed `c` and stream has evaluated its value up to `b` which means up to second element.

The behavior of stream will be more clear in example of creating an infinite stream.

Creation of an infinite length stream

Streams in Scala can be constructed in two ways:

- Using the `#::` operator:

```
scala> import Stream._

import Stream._
```

- The `#::` operator is defined under Scala `Stream` class. Therefore, we have imported `Stream._`.

```
scala> var msc = 1 #:: 2 #:: empty

msc: scala.collection.immutable.Stream[Int] = Stream(1, ?)
```

- We have created a Stream using the `#::` operator. Here empty is just an empty stream. Let us just type empty on console and see what the result is:

```
scala> empty

res23: scala.collection.immutable.Stream[Nothing] = Stream()
```

- Now we can see that empty will just create an empty stream.

- Using `cons ()` function to create an infinite stream:

```
cons(anObject, streamOfObjects)
Constructor cons will take anObject and a streamOfObjects.

scala> var ms  = cons(1, empty)

ms: Stream.Cons[Int] = Stream(1, ?)
```

We have seen that empty is also an empty stream. So, cons is taking 1 as `anObject` and empty as a `streamOfObject`:

```
def strFun(n : Int) : Stream[Int] = {
     println("Evaluating next element of Stream")
     cons(n, strFun(n+1))
     }
```

The `strFun` function will calculate an infinite stream, and every time it will evaluate a stream member, it will print message `Evaluating next element of Stream` before evaluation of the next number:

```
scala> var infStr = strFun(1)

Evaluating next element of Stream
infStr: Stream[Int] = Stream(1, ?)
```

Calling the `strFun` function with argument 1 has evaluated the first element of the stream. It has just evaluated the first element of stream; hence, it has printed the message first and then the stream:

```
scala> infStr(0)

res31: Int = 1
```

Since the first element has already been evaluated, the indexing of the first element has not printed the message:

```
scala> infStr(1)

Evaluating next element of Stream
res32: Int = 2
```

Indexing the second element of the stream has again printed the message first and then the second element of stream. It is concluding that in order to get the second element, strFun is used to evaluate the second element of stream because it has not been used before.

Let us index again the second element of the stream and explore the result:

```
scala> infStr(1)

res33: Int = 2
```

To our surprise, the message has not been printed. It says that it is using the previously calculated value of the second element. And, it is now very clear that the value has been memoized so it is used from memoized location.

Stream is immutable

We cannot change individual elements of a stream because they are immutable. We can observe the immutability of streams in the following example:

```
scala> infStr(1)

res33: Int = 2
```

Now let us assign some other value to this Stream and observe the output:

```
scala> infStr(1) = 10

<console>:17: error: value update is not a member of Stream[Int]
               infStr(1) = 10
                    ^
```

Assigning a new value to a stream results in error. This is because streams are immutable.

Creating a stream from another

Many operations on one stream will create another. The operation like map on one stream will create another stream:

```
scala> var minfStr = infStr.map(x => x*x)

minfStr: scala.collection.immutable.Stream[Int] = Stream(1, ?)
```

The map function is mapping each element of stream to its square. But new variable minfStr is also a stream. Only the first value of minfStr has been realized as clear from code result:

```
scala> minfStr(2)
Evaluating next element of Stream
Evaluating next element of Stream
res0: Int = 9

scala> minfStr(2)

res1: Int = 9
```

Here, in order to evaluate the elements of minfStr, at first the element of infStr is getting evaluated. After getting the realized value of infStr, the map function is working. But, is this clear that the output of the map function is memoized or not? We will explore this in the following part:

```
def mapStr( x : Int) : Int = {
    println("Mapping Values")
    x*x
}

scala> var infStr = strFun(1)

Evaluating next element of Stream
infStr: Stream[Int] = Stream(1, ?)


scala> var minfStr = infStr.map(x => mapStr(x))

Mapping Values
minfStr: scala.collection.immutable.Stream[Int] = Stream(1, ?)


scala> minfStr(2)
```

```
Evaluating next element of Stream
Mapping Values
Evaluating next element of Stream
Mapping Values
res5: Int = 9


scala> minfStr(2)

res6: Int = 9
```

Stream to list

A stream can be changed to a list. In order to explore this, let us first start with changing our infinite stream `infStr` to list and understanding the result:

```
scala> var ls = infStr.toList   (Caution)

Evaluating next element of Stream
Evaluating next element of Stream
Evaluating next element of Stream
Evaluating next element of Stream
```

This will start generating an infinite list, so be cautious. We will explore other properties of a stream using a finite Stream:

```
scala> var strm = Stream(1,2,3,4,5)

strm: scala.collection.immutable.Stream[Int] = Stream(1, ?)
```

We have created a simple stream of five elements:

```
scala> var ls  =  strm.toList

ls: List[Int] = List(1, 2, 3, 4, 5)
```

We have changed the stream to a list using the `toList` function.

Appending one stream to another

A stream can be appended to another stream. Let us explore this in the following example:

```
scala> var strm = Stream(1,2,3,4,5)

strm: scala.collection.immutable.Stream[Int] = Stream(1, ?)
```

We have created a stream `strm` with five integers:

```
scala> var strm1 =  Stream(6,7,8)

strm1: scala.collection.immutable.Stream[Int] = Stream(6, ?)
```

We have created another stream with another three integers:

```
scala> var astrm = strm.append(strm1)

astrm: scala.collection.immutable.Stream[Int] = Stream(1, ?)
```

The method `append` of the `Stream` class will append one Stream to another.

Length of a stream

The `Stream` class consists of a method length, which calculates the length of a stream.

```
scala> astrm.length

res1: Int = 8
```

Treating length function with infinite stream is illogical. It will start evaluating each element of that stream and you will never get the length.

Some mathematical functions of the stream class

Methods sum, max, and min methods will calculate summation, maximum, and minimum elements in a stream, respectively:

```
scala> astrm.sum

res5: Int = 36

scala> astrm.max

res2: Int = 8
```

```
scala> astrm.min

res3: Int = 1
```

Some more methods of the stream class

In the following part of section, we are going to explore some more functions, that can be applied on stream:

```
scala> var strm = Stream(1,2,3,4,5)
scala> strm.count(x => x%2==0)

res6: Int = 2
```

In the preceding code, the count method finds the total number of elements in stream, which are divisible by 2:

```
scala> strm

res7: scala.collection.immutable.Stream[Int] = Stream(1, 2, 3, 4, 5)

scala> strm.find(x => x >3)

res10: Option[Int] = Some(4)
```

Streams (lazy sequence) in Clojure

Streams in Clojure can be created using macro lazy-seq. The elements of sequence created by lazy-seq are evaluated lazily.

Let us understand creation of a lazy sequence in Clojure using following code example:

```
(defn strFun [n]
( lazy-seq  (println "Evaluating next element of Stream") (cons n (strFun
(inc n)))))
```

Clojure function strFun creates a lazy sequence of integers, starting from the integer argument of strFun. Before evaluating any member of a lazy sequence, the strFun function will print message, Evaluating next element of Stream:

```
user=> (take 3  (strFun 1))

Evaluating next element of Stream
Evaluating next element of Stream
Evaluating next element of Stream
(1 2 3)
```

We can see how elements of a lazy sequence are created.

Creating a memoized function of lazy sequences in Clojure

In the following code, we are going to create a memoized function in Clojure:

```
user=> (defn strFun [n]
  #_=> ( lazy-seq  (println "Evaluating next element of Stream") (cons n
(strFun (inc n))))))

#'user/strFun
```

In order to make a function memoized, we have to use Clojure keyword memoize as follows:

```
user=> (def memzstrFun (memoize strFun))
#'user/memzstrFun

Function strFun is memoized.

user=> (take 3 (memzstrFun 1))

Evaluating next element of Stream
Evaluating next element of Stream
Evaluating next element of Stream
(1 2 3)

user=> (take 3 (memzstrFun 1))
(1 2 3)
```

Some algorithms on stream

We have found that streams are infinite sequence. Hence, streams are used to create mathematical sequences. We find many mathematical series, which are used in day-to-day simulations and mathematical modeling. Some of mathematical series are as follows:

- Arithmetic progression
- Geometric progression
- Harmonic progression
- Fibonacci series

Brownian motion path

Let us explore some mathematical series using lazy sequences. I should start with the Arithmetic series.

Arithmetic progression

Arithmetic progression is a mathematical sequence where the difference between two consecutive elements is constant:

2, 5, 8, 11, 14, 17,...

The preceding mathematical sequence is an arithmetic progression, and the difference between any two consecutive elements is three. This constant difference is known as **common difference**. First term of the series is known as **initial term**. If *1* is the initial term of an arithmetic progression, then the nth term a_n is calculated as follows:

$a_n = a_1 + (n\text{-}1)^*d$

Where *d* is common difference.

Arithmetic progression in Scala

In this section, we will implement the arithmetic progression using Scala `Stream`:

```
Scala>  import Stream._
scala> def arithmeticProgression(ft : Double, cd: Double ) : Stream[Double]
= {
                cons(ft, arithmeticProgression(ft+cd, cd ))
        }

arithmeticProgression: (ft: Double, cd: Double) Stream[Double]
```

The `arithmeticProgression` method is a recursive function to calculate arithmetic progression. It takes two arguments: `ft`, which is the first term, and `cd`, which is common difference. The `arithmeticProgression` method returns a stream of arithmetic progression:

```
scala> val ap = arithmeticProgression(2,3)

ap: Stream[Double] = Stream(2.0, ?)
```

We have created a Stream of arithmetic progression with initial term 2 and common difference 3:

```
scala> print(ap.take(5).toList)

List(2.0, 5.0, 8.0, 11.0, 14.0)
```

The preceding code line creates an arithmetic progression of length 5, which is transformed into a list and finally printed.

Arithmetic progression in Clojure

In this section, we will implement the creation of an arithmetic progression in Closure using macro `lazy-seq`:

```
(defn arithmeticProgression [ft cd]
  ( lazy-seq  (cons ft (arithmeticProgression (+ ft cd) cd)))))
```

Clojure function `arithmeticProgression` will take initial term as `ft` and common difference as `cd` and return a lazy sequence of arithmetic progression:

```
user=> (println (take 10 (arithmeticProgression 2 3)))

(2 5 8 11 14 17 20 23 26 29)
```

Finally, first 10 members of arithmetic progression starting with 2 with common difference of three, are printed out.

Standard Brownian motion

Standard Brownian motion is a stochastic process, Wt for $0 \leq t$ with the following characteristics:

- $W_0 = 0$ (with probability 1).
- W_t has continuous path $t \rightarrow W_t$.
- W_t has an independent increment. This means $W_{t+s} - W_s$ will be independent of $(W_r$ where $r \leq s$).
- W_t has Gaussian increment. This means $W_{t+s} - W_s$ is normally distributed with mean 0 and variance t.

Standard Brownian motion is also known as the **Wiener process**. Many applications of the Wiener process can be found in financial simulations, economics, and physics.

The algorithm to create a standard Brownian motion is as follows:

1. Set $W_1 = 0$
2. Loop $W_{t+1} = W_t + N(0,1)$

Standard Brownian motion in Scala

In the following, we will generate a standard Brownian motion path:

```
import Stream._
def brownianPathGenerator( x : Double) : Stream[Double] = {
        val y = scala.util.Random.nextGaussian
        cons(x, brownianPathGenerator(x+y))
    }
```

Scala function `brownianPathGenerator` will create a Scala stream of Wiener process. In order to generate a Gaussian number of mean 0 and standard deviation 1, we will use `scala.util.Random.nextGaussian`:

```
scala> val brownianPath = brownianPathGenerator( 0 ).take(10)
brownianPath: scala.collection.immutable.Stream[Double] = Stream(0.0, ?)
```

```
scala> brownianPath.toList.foreach(println)
0.0
-0.8884224650658452
-1.1674519715899003
-1.1101494700449688
-2.747288986100342
-1.8088902191538734
-3.1200362823554455
-1.8655095303265137
-2.9850674353166182
-2.997584837394881
```

We can see that we have created 10 elements of standard Brownian path.

Standard Brownian motion in Clojure

Following is the implementation of a Clojure function to generate standard Brownian path lazily:

```
(defn brownianPathGenerator [x] ;; Line One
  (def rndm (java.util.Random.)) ;; Line Two
  (def y (.nextGaussian rndm)) ;; Line Three
  (lazy-seq (cons x (brownianPathGenerator (+ x y))))) ;; Line Four

#'user/brownianPathGenerator
```

The nextGaussian function of the java.util.Random class will create a Gaussian random number of mean 0 and standard deviation 1. The Clojure function brownianPathGenerator will generate Standard Brownian path lazily. *Line Two* of the code defines the rndm variable as java.util.Random. In *Line Three,* one Gaussian number of mean 0 and standard deviation 1 is being generated and set in y. *Line Three* is obviously adding a new Gaussian number and being recursive to call the brownianPathGenerator for the next element in a lazy sequence:

```
user=> (take 10 (brownianPathGenerator 0))
(0 1.244783580769308 0.29522144078955803 0.9899565158452042
2.2607158854920852 2.686127999299895 3.26925523900515 3.9852861429115034
4.718989955572183 4.391362795823513)
```

The preceding line of code takes 10 points of a Standard Brownian path.

Fibonacci series

Fibonacci series is one of the most celebrated series of integers. There are two types of Fibonacci series, depending of its starting integer. I have found best place to read about Fibonacci series is Wikipedia. Let me discuss the two forms one by one.

First form of Fibonacci series

Let us discuss, first form of Fibonacci.

We will start by saying:

$F_1 = 0$

$F2 = 1$

And other elements of series are defined by recurrence relation:

$F_{n+1} = F_n + F_{n-1}$

Where F_n is n^{th} element of Fibonacci series.

Second form of Fibonacci series

Following is second form of Fibonacci. First element of second form is 1:

$F1 = 1$

$F2 = 1$

And other elements of series are defined by recurrence relation

$F_{n+1} = F_n + F_{n-1}$

Where F_n is n^{th} element of Fibonacci series.

Fibonacci series in Scala

Now it is time to explore Fibonacci series in Scala. It can be written in many ways. One way to write it is as follows:

```
import Stream._
def fiboSeries(x : BigInt, y : BigInt) : Stream[BigInt] = {
    cons(x,fiboSeries(y,x+y))
  }

fiboSeries: (x: BigInt, y: BigInt)Stream[BigInt]
```

Scala function `fiboSeries` will take two Scala `BigInt` as two starting consecutive elements of Fibonacci series and return a stream of Fibonacci series. Why `BigInt`? This is because Fibonacci elements become very large in magnitude:

```
scala> val fibSeriesWithZero = fiboSeries(0,1)
fibSeriesWithZero: Stream[BigInt] = Stream(0, ?)
```

Using the `fiboSeries` function, and 0 and 1 as the first and second elements of series, respectively, we are evaluating the series as stream:

```
scala> fibSeriesWithZero.take(5).foreach(println)
0
1
1
2
3
```

First five elements of Fibonacci series are printed:

```
scala> val fibSeriesWithOne = fiboSeries(1,1)

fibSeriesWithOne: Stream[BigInt] = Stream(1, ?)
```

Using the `fiboSeries` function, and 1 and 1 as the first and second elements of series, we evaluate the series as a stream:

```
scala> fibSeriesWithOne.take(5).foreach(println)
1
1
2
3
5
```

The first five values of Fibonacci series are printed.

Fibonacci series in Clojure

In Clojure Fibonacci can be implemented in following way:

```
(defn fiboSeries [x y]
(lazy-seq (cons x (fiboSeries y (+ x y)))))

#'user/fiboSeries
```

Similar to Scala function `fiboSeries`, the closure function `fiboSeries` will calculate the lazy sequence of Fibonacci series. It is also a recursive function:

```
user=> (take 5 (fiboSeries 0 1))

(0 1 1 2 3)
```

We take out five elements of Fibonacci series, starting with integers 0 and 1:

```
user=> (take 5 (fiboSeries 1 1))
(1 1 2 3 5)
```

In code line above, we are using `fiboSeries` function to calculate the sequence of five Fibonacci elements, starting from 1 and 1.

Summary

Lazy evaluation delays the evaluation of an expression till the value of expression is needed. Delaying the expression evaluation speeds up the program in many cases, and optimizes the memory utilization.

Memoized functions speed up in their subsequent calls as their output value in the first call is memoized.

Streams are lazy list. An element of a stream is evaluated on-demand and value is memoized. In subsequent calls on stream object for same element, it returns memoized value..

10
Being Lazy - Queues and Deques

In `Chapter 8`, *Queues* we looked at functional queues. Deques are a kind of enhanced queues, allowing you to perform insertion and deletion at both the the front and rear ends of the queues.

Why would we need to insert and delete elements from both ends? Well, this is how we would check whether a string is a palindrome. A palindrome is a string that reads the same when read backwards, that is, `s == s.reverse` always holds for palindromes. For example, *MADAM* is a palindrome. (See `http://www.fun-with-words.com/palin_example.html` for more fun examples of palindromes.)

The following figure shows a string and how we could use deletion at both ends to check whether it is a palindrome:

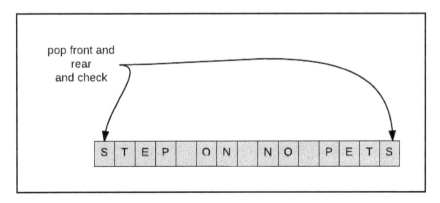

The algorithm is pretty simple. We treat the string as a *deque* of characters. We keep removing successive elements from both the front and rear ends of the string and compare them. If we find a mismatch, the string is not a palindrome.

Another example is how we would model a popular restaurant serving its customers on a busy Sunday. In such situations, people queue up and wait for their turn. When someone's turn comes, the host would ask the customer whether they would be okay sitting in a booth. If the customer is okay, he or she will be removed from the rear and shown to the booth.

On the other hand, if the customer rather prefers a table, the host would request the customer to wait if no tables are available. In this case, the customer would be temporarily removed from the queue, and the next person in the queue would be served; the waiting customer would be enqueued at the front again.

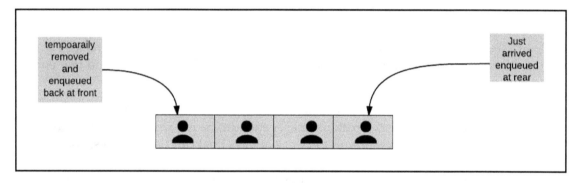

In this chapter, you will learn about functional dequeues. We will take a quick look at how deques are implemented imperatively. Next, we will look at amortization, an algorithm technique that underlies this implementation.

We will use deferred (also known as lazy) evaluation for the implementation. A refresher on lazy evaluations will help. So we will play a bit with `Stream`, that is, Scala's lazy lists.

Lastly, we will implement lazy queues first, improving on the queue implementation we presented in `Chapter 8`, *Queues*. Finally, we will implement dequeues using similar techniques.

By the end of this chapter, you will understand how lazy evaluation factors into functional algorithm design.

Imperative implementations

Implementing a deque is simple in the imperative world. We could use a doubly linked list that will give us $O(1)$push and pop at both ends.

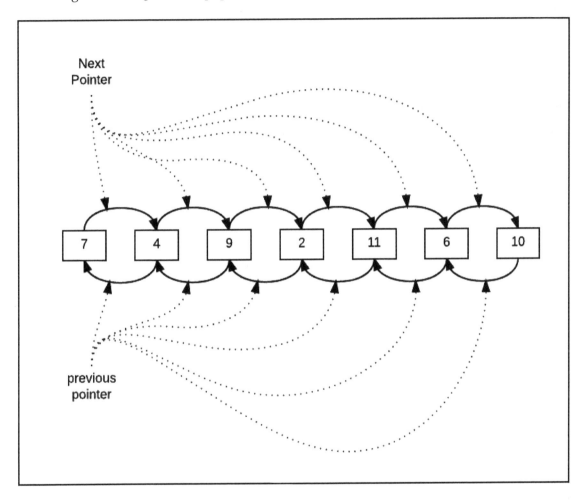

If we keep track of the first and the last elements, adding or removing elements at either end would just be a matter of twiddling a few pointers. Of course, we cannot use this model as we never ever mutate a data structure in place. What about efficiency then?

Hang on! We will soon see how *amortization* and *laziness* help us design an efficient deque algorithm.

Amortization

To better understand the concept of amortization, let's look at a dynamic array. This is an array that would grow if there is no space to add a new element. We could do this as follows:

1. Allocate a new array double the size of the current array.
2. Copy all the elements from the current array to the new array.
3. Make the new array the current array.

Here is a sample run of the algorithm depicted pictorially:

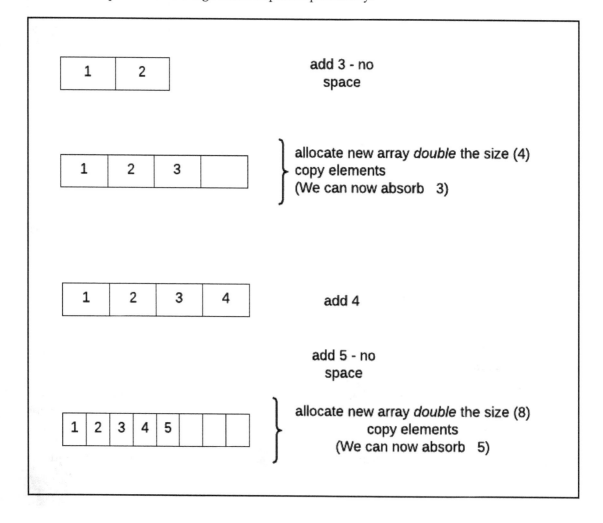

This allocation and copying obviously incur $O(n)$ cost once in a while. If most of the elements incur a $O(1)$ cost, we should be fine though.

If you continue to trace the growth of this array, you will soon realize that the allocate/copy operations occur less frequently as the number of elements grow.

Most of the insert operations would have $O(1)$ complexity. In other words, an insertion would complete in *amortized constant time*.

Problem with queues

Consider our queue implementation using two lists from the previous chapter. When the out list becomes empty, we substitute it with the reversed in list.

To jog your memory, here is the relevant code snippet:

```scala
scala> def pop(queue: Fifo): (Int, Fifo) = {
     |    queue.out match {
     |      case Nil => throw new IllegalArgumentException("Empty queue");
     |      case x :: Nil => (x, queue.copy(out = queue.in.reverse, Nil))
     |      case y :: ys => (y, queue.copy(out = ys))
     |    }
     | }
pop: (queue: Fifo)(Int, Fifo)
```

Note the second case clause where the substitution happens. When we have a large number of insertions, reversing the in list would possibly incur $O(n)$ cost. This would happen once in a while, but that would still be something at least.

So most of our push and pop operations would be $O(1)$, given the occasional in list reversal that could be $O(n)$. This is the concept of amortization, where the occasional cost of reversal is paid off by having $O(1)$ push and pop operations.

We may have an occasional expensive operation; however, the overall pop cost will be $O(1)$.

Strict versus lazy

Consider the operation of zipping up two lists. The `zip` method pairs elements from the first list with the elements from the second list. Here is a sample run:

```
scala> List(1,2,3).zip(List(4,5,6,7))
res8: List[(Int, Int)] = List((1,4), (2,5), (3,6))

scala> List(1,2,3).zip(Nil)
res9: List[(Int, Nothing)] = List()
```

All the elements of both the lists are visited to create a zipped list. The following figure shows the `zip` operation in action:

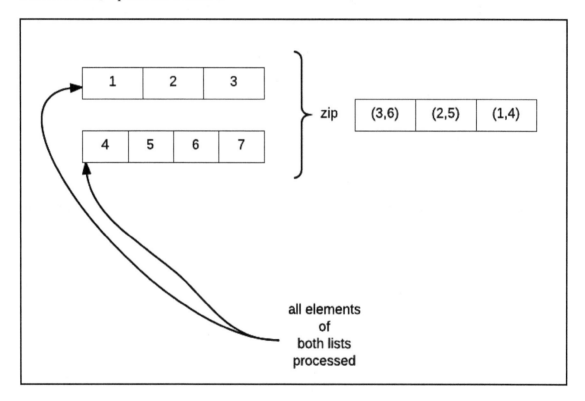

As another example of strict evaluation, consider the `reverse` method of `List`:

```
scala> List(1,2,3).reverse
res11: List[Int] = List(3, 2, 1)
```

The `reverse` method also visits all the elements of the list. On the other hand, consider the following:

```
scala> val q = List.range(1, 1000000).view.reverse
q: scala.collection.SeqView[Int,List[Int]] = SeqViewR(...)

scala> (q take 10) foreach println
999999
999998
999997
. . .
```

Only the first 10 elements of the lazy lists are computed *on demand*.

Streams

Scala's streams are lazy lists, an *infinite* sequence of elements. For example, just like we zip two lists together, we could zip two streams too:

```
scala> def zip(p: Stream[Int], q: Stream[Int]): Stream[(Int, Int)] =
     | (p.head, q.head) #:: zip(p.tail, q.tail)
zip: (p: Stream[Int], q: Stream[Int])Stream[(Int, Int)]

scala> val r = zip(Stream.from(9), Stream.from(10))
r: Stream[(Int, Int)] = Stream((9,10), ?)
```

As you can see, the type of r variable is another stream. Also, the stream has a `head` element, a pair `(9,10)`, and a function (also called a thunk). The function (in this case, `zip`) is called to produce successive elements.

As you can see, the `tail` is a question, meaning it is not yet computed; therefore, it is not yet known.

Why are streams called lazy? Note the definition of `zip`, which is defined recursively. There is no condition related to termination; the calls just recur. A recursive call is not evaluated right away but only when it is needed.

The idea is *to force* a call to the thunk function when we need the value(s).

Scala's call by name parameters are evaluated if needed:

```
scala> def evenOrOdd(even: Boolean, i: => Int) = {
     |     if (even) {
     |       2
     |     } else {
```

```
        |        i * 2 + 1
        |     }
        |   }
      evenOrOdd: (even: Boolean, i: => Int)Int

scala> evenOrOdd(true, { println("You there"); 1 })
res8: Int = 2

scala> evenOrOdd(false, { println("You there"); 1 })
You there
res9: Int = 3
```

When the first argument is `true`, the parameter `i` is never evaluated; therefore, we don't see the "You there" message. See `http://daily-scala.blogspot.in/2009/12/by-name-parameter-to-function.html` for more information on call by name.

For example, when we need to print values, the same number of elements are actually produced by calling the function:

```
scala> (r take 10) foreach println
(9,10)
(10,11)
(11,12)
...
```

The following diagram shows how elements are produced on demand and memoized:

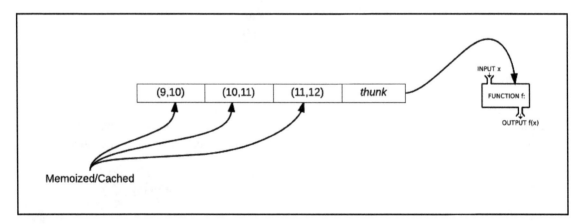

Another important feature of streams is that the values already computed are *memoized*. Memoization is used to avoid calling the thunk function again and again; instead, using it, the value is saved and returned when requested.

You guessed right! This is a space/time trade-off once again; we trade some memory to save time that is otherwise spent in recomputing the value.

As shown in the preceding diagram, the stream caches the pairs already computed.

Streams meet queues

Going back to our functional FIFO queue implementation, do you remember we had to pay the price of occasional reversal? The invariant asserted that the `out` list would never be empty when the `in` list is non-empty.

In case the invariant is violated, we can restore it by *reversing* the `in` list and turning it into a new `out` list. We could exploit this laziness to address the once in a while (and possibly costly) reversal. The idea is to make the `out` list a `Stream`. The `in` list remains a strict list, as before. This helps us do the reversal *on demand*:

```
    case class LazyQueue(out: Stream[Int], outLen: Int, in: List[Int],
inLen: Int) {

    def push(elem: Int) = {
      val p = makeLazyQueue(out, outLen, elem :: in, inLen + 1)
      println(s"pushed: ${elem} - ${p}")
      p
    }

    def pop: (Int, LazyQueue) = {
      val q = (out.head, makeLazyQueue(out.tail, outLen - 1, in, inLen))
      println(s"popped: ${q._1} and ${q._2}")
      q
    }

    def empty = out.isEmpty && in.isEmpty
    }
```

Both the `push` and `pop` methods have `println` statements to help you trace the flow of the execution.

Note the invariant where the `out` stream is always larger than the `in` list. The invariant is violated, for example, when the `in` list gets bigger. We restore it by converting the *reversed* `in` list into a `Stream` and appending it to the `out` stream.

Here is the `makeLazyQueue` method that makes sure the invariant is held after each `push` and `pop` operation:

```
def makeLazyQueue(out: Stream[Int], outLen: Int, in: List[Int], inLen:
Int): LazyQueue = {
    if (inLen <= outLen) {
      LazyQueue(out, outLen, in, inLen)
    } else {
      val newOutStream = copyInToOut(out, in, Stream.empty)
      val newOutSize = outLen + inLen
      val newInList = Nil
      val newInSize = 0

      LazyQueue(newOutStream, newOutSize, newInList, newInSize)
    }
  }
```

The `if (inLen <= outLen) {` condition evaluates to `true` when the invariant holds. We just create a newer version of the queue and return it.

Now check out the following clause:

```
else {
      val newOutStream = copyInToOut(out, in, Stream.empty)
  ...
  }
```

This clause matches when the invariant is violated. In this case, we call the `copyInToOut` method to restore it.

Here is the `copyInToOut` method:

```
def copyInToOut(out: Stream[Int], in: List[Int], revIn: Stream[Int]):
Stream[Int] =
    in match {
    case Nil => Stream.empty
    case x :: xs if out.isEmpty => Stream.cons(x, revIn)
    case x :: xs => Stream.cons(out.head, copyInToOut(out.tail, in.tail,
Stream.cons(x,
        revIn)))
  }
```

This is a very beautiful algorithm. Let's look at the details and see how *lazy evaluation* helps.

Here's the first case clause:

```
case Nil => Stream.empty
```

It matches when the `in` list is empty. If so, there is nothing to be done and we just make the out stream empty too.

Here's the second case clause:

```
case _ :: _ if out isEmpty => if out.isEmpty => Stream.cons(x, revIn)
```

This matches when the `in` list is not empty. Here, the invariant is violated: the `out` stream is empty, while the `in` list is not.

We create a `Stream` with x as the `head`. What is the second argument? It is a stream. Let's look at the third clause and then we will understand more about the second argument.

Here's the third case clause:

```
case x :: xs => Stream.cons(out.head, copyInToOut(out.tail, in.tail,
Stream.cons(x,
      revIn)))
```

This matches when the `out` stream has one element less compared to the `in` list, that is, when this clause matches `outlen == inLen - 1`.

In this case, we create a `Stream` and set its `head` to x, which is the first element of the `in` list.

The `tail` of this stream is a recursive call to `copyInToOut`. When this result, which is also a `Stream`, is accessed, the reversal happens *incrementally*. Here is an figure with an example scenario when we push a sequence of 10, 20, and 30 to the lazy queue:

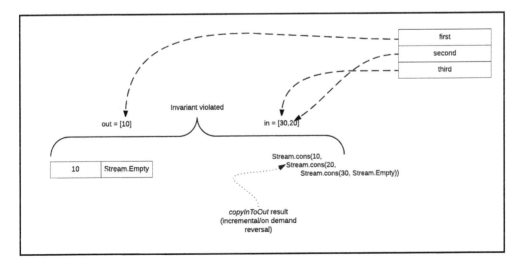

Here is the REPL session using the preceding scenario:

```scala
scala> import com.fpdatastruct.deque.LazyQueue._
import com.fpdatastruct.deque.LazyQueue._

scala>    val emptyQ = makeLazyQueue(Stream.empty, 0, Nil, 0)
emptyQ: com.fpdatastruct.deque.LazyQueue.LazyQueue =
LazyQueue(Stream(),0,List(),0)

scala>    val q = List(10, 20, 30).foldLeft(emptyQ)((acc, x) => acc.push(x))
q: com.fpdatastruct.deque.LazyQueue.LazyQueue = LazyQueue(Stream(10,
?),3,List(),0)

scala>    val (a1, q1) = q.pop
a1: Int = 10
q1: com.fpdatastruct.deque.LazyQueue.LazyQueue = LazyQueue(Stream(20,
?),2,List(),0)

scala>    val (a2, q2) = q1.pop
a2: Int = 20
q2: com.fpdatastruct.deque.LazyQueue.LazyQueue = LazyQueue(Stream(30,
?),1,List(),0)
```

We have seen the use of `foldLeft` for populating a data structure. The final queue is returned as the resulting accumulator value.

A sense of balance

Many data structures have a balance invariant. After every update to the tree, the invariant is restored by rebalancing the structure. Why do we need this balancing? What do we mean by balance?

A Binary Search Tree, for example, could degenerate into a list. For example, consider a scenario where you insert sorted data into a BST. You will get a tree whose nodes have no left children. To all intents and purposes, you have constructed just a linked list in the garb of a tree. This would lead to pathetic access performance for $O(n)$. A balanced BST won't have this problem.

A tree is *perfectly balanced* if the left and right subtrees of any node are of the same height.

We also have *almost perfectly balanced* trees. The subtrees' heights may differ by at most 1.

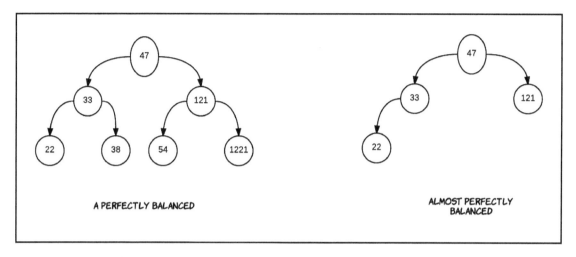

As we will soon see in the next chapter, balancing a BST allows us to have guaranteed *O(logn)* search times. The next chapter discusses Red-Black trees, which are a very popular balanced variant of the BST.

If the updates do not disturb the existing balance too drastically, we could defer the balancing for a while, as we will soon see in the following discussion on deques.

Amortized deques

Coming back to deques, we replace the list in the queue with a stream so both `in` and `out` become `Streams` objects. If we try to keep both the streams *balanced*, we would have an efficient deque implementation.

In this case, by balance, we mean both the streams will have almost the same number of elements. For example, both the streams would be non-empty when the deque contains two or more elements.

Let's say no stream is bigger than the other by a factor, say, `c > 1`. If one stream becomes too long, we move the elements to the other.

Let's look at stream operations a bit more so we could understand the code that follows:

```scala
scala> val s = 1 #:: 2 #:: 3 #:: 4 #:: Stream.empty
s: scala.collection.immutable.Stream[Int] = Stream(1, ?)
```

We define s as a stream:

```scala
scala> val p = s.drop(2)
p: scala.collection.immutable.Stream[Int] = Stream(3, ?)
scala> p foreach println
3
4
```

Calling the `drop(n)` method gives us another stream with the n elements in front removed:

```scala
scala> val p = s.take(2)
p: scala.collection.immutable.Stream[Int] = Stream(1, ?)

scala> p foreach println
1
2
```

The first two elements are copied to the new stream, as the following figure shows:

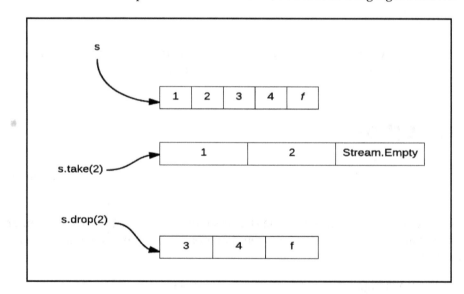

Note the `take` and `drop` methods, which are invoked on the stream, work just like their list counterparts–only lazily, though.

These two methods help us move the elements from one stream to another. Armed with these tidbits, let's look at the deque implementation:

```
  case class Deque(outLen: Int, out: Stream[Int], inLen: Int, in:
Stream[Int], c: Int = 2) {
    def pushFront(elem: Int): Deque = {
      adjustStreams(outLen+1, Stream.cons(elem, out), inLen, in, c)
    }

    def popFront() : (Int, Deque) = {
      out match {
        case Stream.Empty => throw new IllegalArgumentException("Empty
queue")
        case x #:: newOut => (x, adjustStreams(outLen-1, newOut, inLen, in,
c))
      }
    }
  }
```

The `pushFront` method just prepends `elem` to the `out` stream. It calls `adjustStream` to make sure the balance is *restored*.

The `popFront` method returns a tuple of two elements. The first is the element that is removed, and the second is the new deque left after the removal. Here's the first clause:

```
  case Stream.Empty => throw new IllegalArgumentException("Empty queue")
```

This matches if there are no elements in the `out` stream. An exception is thrown in this case.

The second clause matches when the stream is not empty, that is when it has one or more elements. It extracts the `head` of the stream and calls `adjustStream` on the `tail`. Both are then packaged into a tuple, which is returned.

The `adjustStream` method is the `workhorse` method. It *balances* both the `in` and `out` streams. Note the constant `c` defaults to 3:

```
  def adjustStreams(outLen: Int, out: Stream[Int], inLen: Int, in:
Stream[Int], c: Int): Deque = {
    if (outLen > c*inLen+1) {
      val newOutLen = (outLen+inLen)/2
      val newInLen  = outLen + inLen - newOutLen

      val newOut = out.take(newOutLen)
      val newIn = in append out.drop(newInLen).reverse
```

```
        Deque(newOutLen, newOut, newInLen, newIn, c)
    } else if (inLen > c*outLen+1) {
        val newInLen = (outLen+inLen)/2
        val newOutLen = outLen + inLen - newInLen

        val newIn = in.take(newInLen)
        val newOut = out append in.drop(newOutLen).reverse

        Deque(newOutLen, newOut, newInLen, newIn, c)
    } else
        Deque(outLen, out, inLen, in, c)
}
```

Note that the adjustment code is the same for both the cases. Whenever we move the elements, we drop them from the source (that is, longer) stream and then reverse and append it to the shorter stream.

We use the `append` method of the `Stream` class to append one stream to another. Here is an example run:

```
scala> val s1 = 1 #:: 2 #:: Stream.empty
s1: scala.collection.immutable.Stream[Int] = Stream(1, ?)

scala> val s2 = 3 #:: 4 #:: Stream.empty
s2: scala.collection.immutable.Stream[Int] = Stream(3, ?)

scala> val s12 = s1 append s2
s12: scala.collection.immutable.Stream[Int] = Stream(1, ?)

scala> s12 foreach println
1
2
3
4
```

Now, let's try using the `Deque`:

```
scala>    val dq = Deque(0, Stream.Empty, 0, Stream.Empty)
dq: Deque = Deque(0,Stream(),0,Stream(),2)
```

Note that each `Deque` is immutable and persistent:

```
scala>    val dq1 = dq.pushFront(1)
dq1: Deque = Deque(1,Stream(1, ?),0,Stream(),2)
```

The first value 1 got pushed to the front, in the out stream:

```
scala>    val dq2 =  dq1.pushFront(2)
dq2: Deque = Deque(1,Stream(2, ?),1,Stream(1),2)
```

The second value 2 triggered a rebalance and the value 1 got pushed to the in stream. In the previous step, there was only one element, so it was pushed to the front. However, when we insert the second value, there is a rebalance and both the streams get an element each:

```
scala>    val dq3 = dq2.pushFront(3)
dq3: Deque = Deque(2,Stream(3, ?),1,Stream(1),2)
```

The third insertion went to the out stream:

```
scala>    val (x,p) = dq3.popFront()
x: Int = 3
p: Deque = Deque(1,Stream(2, ?),1,Stream(1),2)
scala>    println(x)
3
scala>    val (y,p1) = p.popFront()
y: Int = 2
p1: Deque = Deque(0,Stream(),1,Stream(1),2)
scala>    println(y)
2
```

However, now if you try to pop the last value 1, as it is in the in stream, you will get an exception. Fixing this exception is left as an exercise for you.

Summary

We looked at another common data structure called a deque. A deque is a double-ended queue, so insertions and deletions are possible at both ends. We looked at pushFront, popFront, pushBack, and popBack operations. We also looked at some applications of deques.

Deques have a simple implementation in the imperative world. A doubly linked list could be used for implementing it. A vector could be used too.

We looked at two important concepts: amortization and lazy evaluation. We took a detailed look at Scala's Streams. We saw what a *thunk* is and how it helps us with *delayed evaluation*.

We then used all these concepts to check whether we can create more efficient queues. As we saw in the previous chapter, reversing an in list and appending it to the out list could be costly.

Then we applied the same concepts to look at how to implement persistent and immutable deques. We also saw how balancing a data structure plays a vital role in improving performance.

With all this knowhow, in the next chapter, we will look at a very popular balancing data structure called a Red-Black tree.

11
Red-Black Trees

In the previous chapter, we touched upon concepts such as amortization and balance in data structures. We need to balance the data structure so it does not degenerate. For example, in a Binary Search Tree (BST), if we insert elements that are already sorted, we get a tree, which is a linked list.

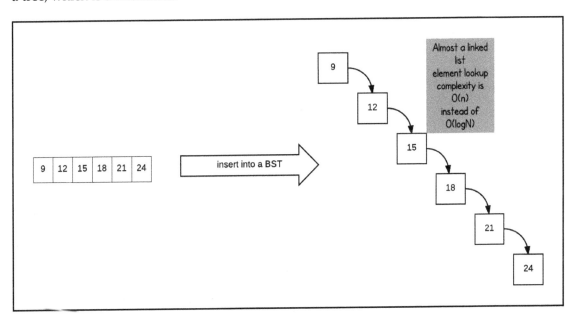

This is highly undesirable if we need a strong guarantee with respect to lookup complexity. Note that the insertion complexity is also $O(n)$.

The solution is to balance the tree so things don't get out of hand. For example, if we could somehow ensure that the *height* of our tree is $O(log_n)$ for any data set, then we will have ensured $O(log_n)$ lookup and insertion/deletion.

Red-Black trees are basically BSTs. However, these trees are unique: every node sports a color, either red or black. This auxiliary information helps us keep the tree balanced, as we will soon see.

Terminology

Let's familiarize ourselves with some terms we'll commonly come across in our upcoming discussion.

Here is a helpful diagram showing the terms:

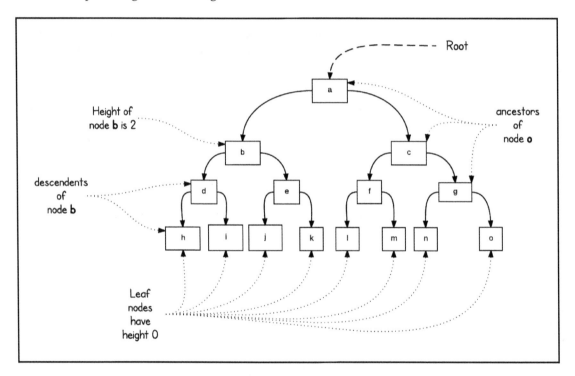

A tree node's height is defined as the number of edges on the longest path to a leaf. A leaf node's height is 0. For example, in the preceding diagram, the height of the node **h** is 0.

The total number of children of a node is collectively referred to as the node's degree. A leaf node's degree is 0. In the preceding diagram, the degree of node a is 2.

A non-leaf node is also called an internal node. For example, in the preceding diagram, node **c** is an internal node, so are **a, b, d, e, f,** and **g**.

Every internal node in the preceding tree has the same degree: 2. This makes this tree a complete binary tree.

Refer to
`http://stackoverflow.com/questions/2603692/what-is-the-difference-between-tree-depth-and-height` for more details and related discussion.

Almost balanced trees

In the tree we just saw, every node's left and right subtrees are of the same height. This makes it a tree that is perfectly height-balanced. However, such trees are very rare; we come across them only when we have large trees with thousands of nodes.

Instead, we could try for trees that are either perfectly height-balanced or somewhere close to that. What do we mean by height-balanced? If the heights of any nodes, left or right subtrees, differ by *at most 1*, it is a height-balanced tree. The complexities of various operations would be almost the same as for a perfectly balanced tree.

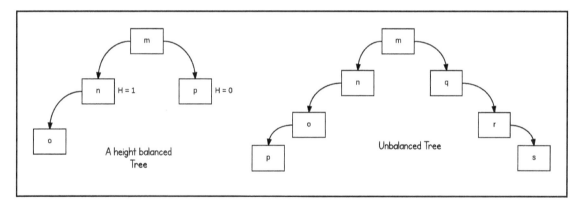

In the preceding diagram, the left tree is height-balanced, whereas the right tree is not. In the left tree, the height of subtrees rooted at **n** is 1. The height of the subtree rooted at **p** is 0. These differ by 1, but we are okay with this little bit of imbalance.

The concept of rotation

Before we jump headlong into the nitty-gritty of Red-Black tree implementation, let's look at a fundamental concept: rotation. Rotations are used in Red-Black trees to restore balance.

Let's look at left rotation first. Rotate the tree counterclockwise so the node **19**, which was earlier a parent of **12**, becomes its right child.

It is always okay to do this as the parent of a node can be made its right child to preserve the BST invariant, namely the right child value should be greater that the parent node. The parent 95 of 12 has now become the parent of **19**. The original left child of **19** was **17**, which is now the right child of **12**. Lastly, **12** is now the new left child of **19**.

Note that we just changed a *fixed* number of pointers. The children of **7** and **17** are not affected at all as also the tree above **95**.

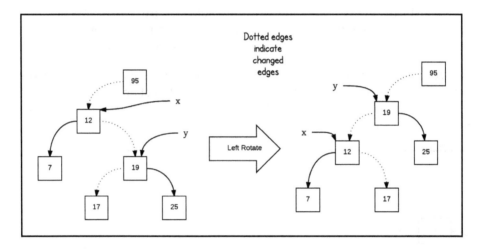

Here is the pseudo code of left rotation:

```
y = x.right
x.right = y.left
if y.left != nil
   then y.left.parent = x
y.parent = x.parent
if x.parent = nil
   then root = y
else if x == x.parent.left
   then x.parent.left = y
else
   x.parent.right = y

y.left = x
x.parent = y
```

The complexity of the rotation is *O(1)*. The change is pretty localized. The right rotation algorithm is similar:

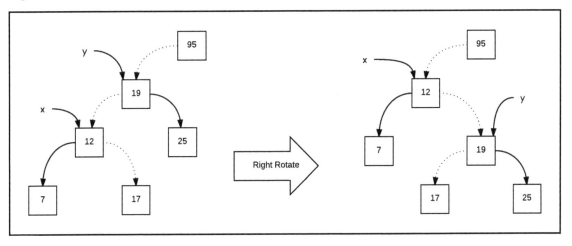

Could you come up with the pseudo code for the right rotation?

Red-Black trees

Given this rotation algorithm, we can now look at the core Red-Black tree.

A Red-Black tree node always has a color, either red or black, with the following invariants:

- A red node can never have a red child
- Every path from the root to an empty node contains the same number of black nodes

An empty node is a leaf nil node. This nil node indicates termination and is also known as a **sentinel node**.

Here is an example of a Red-Black tree. Note that each node is annotated with its *black height*. The black height is the number of black nodes from the node to the leaf.

Note these important points:

1. The root is always black.
2. Every leaf is black.
3. Both the children of a red node are black (*as a red node cannot have a red child*).
4. Every path from a node to a leaf contains the same number of black nodes.

The following diagram shows a Red-Black tree:

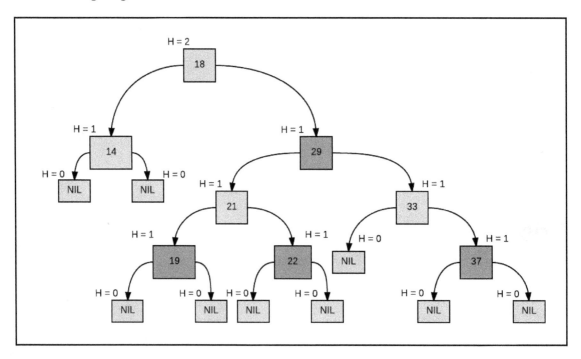

Take each of the invariants from earlier and check whether the tree satisfies each one of them.

When we look at rebalancing, we will see how the first invariant is always assured.

Actually, rebalancing works on restoring the second invariant only, as you will soon see:

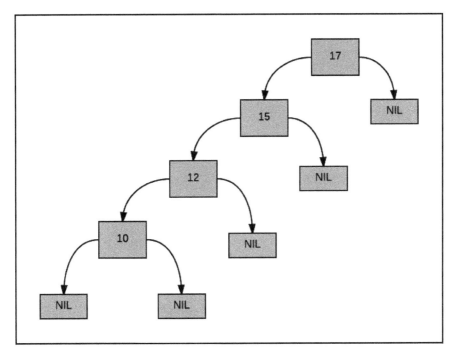

The preceding diagram is not a Red-Black tree. To start with, the first invariant is violated. The number of black nodes on the left path of the root node is four, which does not match with the number of black nodes on the right path, which is one.

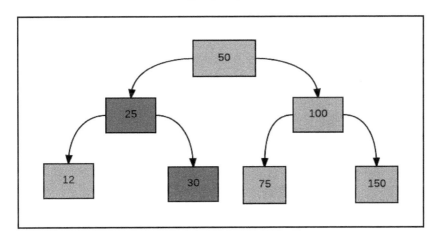

The preceding figure doesn't illustrate a Red-Black tree as well, as node **25** is the parent of node **30** and both are colored red.

Inserting a node

When we insert a node, we color it red. This means that the newly inserted node will never change the black height of any ancestor!

However, it could violate another invariant, namely a red node can never have a red parent. Once we fix this invariant violation, we are done! Here's the code to do this:

```
sealed trait Color
case object Red extends Color
case object Black extends Color
```

We know the advantages of sealing a trait. The `Color` trait can now be used only in this source file. This helps the compiler check for exhaustive matching. As seen earlier, the `List` and `BinTree` traits were also sealed.

The case object idiom creates singleton objects, just like `List.Nil` and `Bintree.Leaf`.

Next, we define our `Tree` class:

```
sealed abstract class Tree {
  def color: Color
}
```

We create two concrete instances of the `Tree` class. The internal nodes are modeled by the case class, that is, `Node`:

```
case class Node(color: Color, left: Tree, value: Int, right: Tree) extends
Tree {
  override def toString = value.toString + " " + left.toString + " " +
right.toString
}
```

The leaf node is modeled as a `case object`. Note how this helps us model it as a sentinel node:

```
case object End extends Tree {
  override def toString = "."
  override val color: Color = Black
}
```

Given all these declarations, here is our `insert` method (the `balance` method will follow):

```
def insert(v: Int, t: Tree): Tree = {
  def ins(s: Tree): Tree = s match {
    case End => Node(Red, End, v, End)
    case node@Node(_, left, value, right) =>
      val root = if (v < value) balance(node.copy(left = ins(left)))
      else if (v > value) balance(node.copy(right = ins(right)))
      else node
      root match {
        case node@Node(Red, _, _, _) => node.copy(color = Black)
        case _ => root
      }
  }
  ins(t)
}
```

The notation `node @ Node(color, left, value, right)` is a feature called **variable binding**. We need to look at the fields of the `Node` object, which a normal pattern match gives us. We also want to bind the matched pattern to a variable. Here's how we can do this:

```
scala> val s = ("one", "two")
s: (String, String) = (one,two)
scala> s match {
     |   case t@(_,_) => println(t)
     |   case _ => println("Huh?")
     | }
(one,two)
```

This helps us refer to the node easily in the else clause when the value is found in the tree.

 Note that the `ins` method is hidden within the `insert` method. The ins method is the real workhorse, which is called by the `insert` method.

This method is similar to a BST node insertion process. If a value is already present, the node holding it is returned. Otherwise, we will keep looking for the right place to insert the value.

Note that this value always replaces a leaf node. This is what `case End => Node(Red, End, v, End)` denotes. When we hit the leaf, we create a new node, as shown in the following figure:

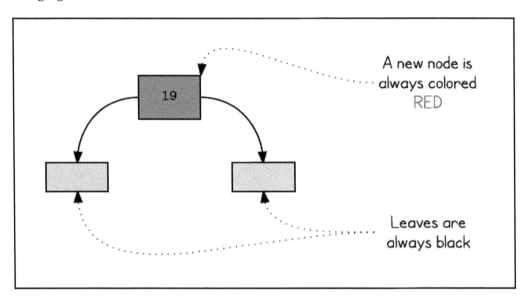

Once the new node is created and returned, we need to make sure the second invariant–a red node cannot be a parent of another red node–is restored.

This is handled by the `balance` method. The subtree returned by the `balance` method is always a valid Red-Black tree.

Before we look at balance, let's learn a bit more about constructor patterns in Scala's pattern matching mechanism.

Here is an illustration:

```scala
scala> case class L1(v1: Int, v2: Int)
defined class L1

scala> case class L2(v3: Int, p: L1)
defined class L2

scala> case class L3(v4: Int, q: L2)
defined class L3

scala> val l1 = L1(10, 20)
l1: L1 = L1(10,20)
```

```
scala> val l2 = L2(30, l1)
l2: L2 = L2(30,L1(10,20))

scala> val l3 = L3(40, l2)
l3: L3 = L3(40,L2(30,L1(10,20)))

scala> l3 match {
     |     case L3(a, L2(b, L1(c, d))) => println(s"${a}, ${b}, ${c}, ${d}")
     | }
40, 30, 10, 20
```

In this example, L1, L2, and L3 are case classes. A constructor pattern matches whether the matched object is an instance of the said case class. As L2 nests within L1 and L3 nests within L2, *deep matching* kicks in.

Note the following statement:

```
case L3(a, L2(b, L1(c, d))) => println(s"${a}, ${b}, ${c}, ${d}")
```

Here, we are matching the L2 object within L1 and L3 object within L2 at the same time. We can use deep matching to arbitrarily go deeper into an object.

Finally, note that the insertion method always colors its root node black. When all of the rebalancing is done, the following statement will make sure the root will always be black:

```
root match {
        case node@Node(Red, _, _, _) => node.copy(color = Black)
        case _ => root
    }
```

Now let's take a close look at the meat of the code, the core tree rebalancing algorithm.

The Black-Red-Red path

The invariant (a red node never has a red child) is violated only when the new node is a child of a red node. In other words, this means we get a black node path that has a red child which in turn has a red child. There are four cases, and we will discuss each one of these cases.

We will develop the balancing method step by step:

```
def balance(color: Color, l: Tree, v: Int, r: Tree): Tree =
    (color, l, v, r) match {
        // We will discuss the four cases, one by one
```

Left, left – red child and grand child

The first case is when a *black* parent has a left *red* child which in turn has a left red child. The pattern uses deep matching to extract the necessary fields and perform a rotation:

```
case Node(Black, gParent@Node(Red, parent@Node(Red, ggChild1, _,
ggChild2), _, gChild), _, child) =>
    gParent.copy(color = Red,
        left = parent.copy(color = Black),
        right = ggParent.copy(color = Black, left = gChild, right = child))
```

Here is the transformation:

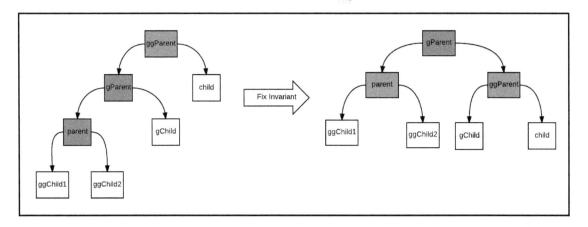

Remember our discussion on tree rotation? The left side tree is rotated around **gParent** so that it becomes a root of the new subtree. Also, the color of **gParent** is changed to red. Note that this recoloring may trigger another black, red, red violation up the tree.

Also note that we make both **parent** and **ggParent** black. This ensures that the first invariant is not violated for any ancestors.

The preceding diagram correlates the transformation with the code:

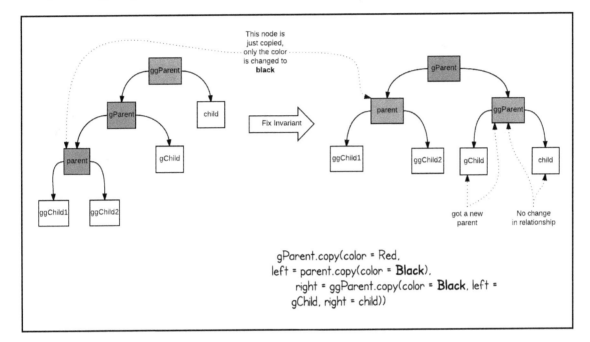

Here are the changes:

- The **parent** node is not changed structurally; only its color is changed from red to black
- The right child of **gParent** is changed to **ggParent**
- The **gChild** becomes the left child of **ggParent**

Note that the right tree is balanced compared to the one on the left

Note how we use the `copy` method given to us by the `case` classes. We do not look at the values at all; instead, we copy the nodes and note the changes explicitly.

The diagram includes the transformation statement so you can correlate it with the structural changes for easier understanding.

Left child, right grand child

This case occurs when the grand child is the right child and not the left child:

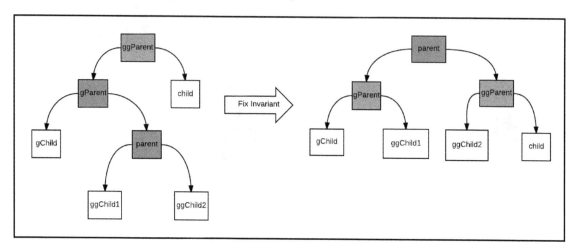

The following case statement reflects this:

```
case Node(Black, gParent@Node(Red, gChild, _, parent@Node(Red, ggChild1, _,
ggChild2)), _, child) =>
    parent.copy(color = Red,
        left = gParent.copy(color = Black, right = ggChild1),
        right = ggParent.copy(color = Black, left = ggChild2, right =
child))
```

Here are the changes:

- The **parent** bubbles up and becomes the subtree root. The original subtree root, **ggParent**, becomes the right child of **parent**.
- The original right child of parent (by the way, this is the rightmost child of the original subtree) becomes the left child of **ggParent**.
- The original left child of parent, **gChild1**, becomes the **right** child of **gParent**.

Drawing the original tree on the left and walking through these steps to derive a right-side-balanced tree will help you understand what is rotation.

Attempting the following quiz will help in understanding the algorithm:

- Why are we always able to make **ggParent** a right child of **parent**?
- Why is it *always* valid to make **gChild1** the right child of **gParent**?

Right child, right grand child

This case is similar to the left-left case we discussed earlier. We rotate around the **gParent** node and the right side of the **gParent** node is not touched at all, except that the parent node's color is changed to black.

All other changes happen on the left side:

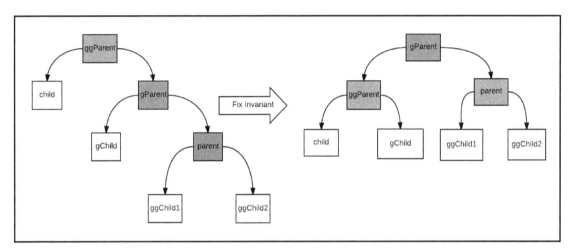

Here is the code that handles this scenario:

```
case Node(Black, child, _, gParent@Node(Red, gChild, _,
parent@Node(Red, ggChild1,
    _, ggChild2))) => gParent.copy(color = Red,
    left = ggParent.copy(color = Black, left = child,
        right = parent.copy(color = Black)))
```

See how the right child is copied with just the color changed to black.

Here are the changes:

1. The **gParent** node becomes the new root of the subtree and changes its left child to **ggParent**.
2. The original left child of **gParent**, **gChild**, becomes the right child of **ggParent**.
3. The color of the parent node becomes black.

Here is another quiz for you:

- Why is it always valid to make **ggParent** a left child of **gParent**?
- Why can't we make **ggChild1** the right child of **ggParent**?

Right, left

This case is like the second one. The violation path is a black node. Its right child is red, whose left child is, in turn, red:

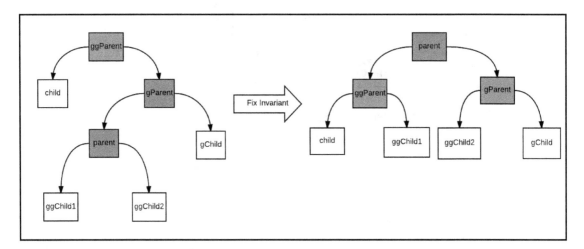

Here is the relevant code snippet:

```
case Node(Black, child, _, gParent@Node(Red, parent@Node(Red, ggChild1, _,
    ggChild2), _, gChild)) =>
        parent.copy(color = Red, left = ggParent.copy(color = Black,
ggChild1),
            right = gParent.copy(color = Black, left = ggChild2))
```

The changes are:

1. The **parent** node becomes the new root of the balanced subtree and keeps its red color. This could trigger rebalancing again by moving upward.
2. The **ggParent** (which was the original root of the subtree) becomes the parent's new left child. The **ggChild1** node becomes the right child of **ggParent**.
3. The **gParent** node goes as the right child of the **parent** node. The node **ggChild2** becomes its left child.

Here is another quiz for you:

- Why are steps 2 and 3 always valid?

Here is the complete rebalancing method:

```
def balance(ggParent: Node): Tree =
  ggParent match {
    case Node(Black, gParent@Node(Red, parent@Node(Red, ggChild1, _,
ggChild2), _, gChild), _, child) =>
      gParent.copy(color = Red,
        left = parent.copy(color = Black),
        right = ggParent.copy(color = Black, left = gChild, right = child))
    case Node(Black, gParent@Node(Red, gChild, _, parent@Node(Red,
ggChild1, _, ggChild2)), _, child) =>
      parent.copy(color = Red,
        left = gParent.copy(color = Black, right = ggChild1),
        right = ggParent.copy(color = Black, left = ggChild2, right =
child))
    case Node(Black, child, _, gParent@Node(Red, gChild, _,
parent@Node(Red, ggChild1,
      _, ggChild2))) => gParent.copy(color = Red,
        left = ggParent.copy(color = Black, left = child,
        right = parent.copy(color = Black)))
    case Node(Black, child, _, gParent@Node(Red, parent@Node(Red, ggChild1,
_,
      ggChild2), _, gChild)) =>
      parent.copy(color = Red, left = ggParent.copy(color = Black,
ggChild1),
        right = gParent.copy(color = Black, left = ggChild2))
    case _ => ggParent
  }
```

This code just puts together all the preceding four cases.

Hint for the quizzes: Think in-order traversal! Rotation needs to make sure the traversal order remains the same.

Note the *catch all* case as well: as there is no color violation, there is no re-balancing to do, and as a result, the tree will be returned as is.

Verifying the transformation

In-order traversal of a BST is a recursive traversal algorithm. It visits the BST tree nodes in a predefined order and always prefers to visit the left child first, then the parent itself, and finally the right child.

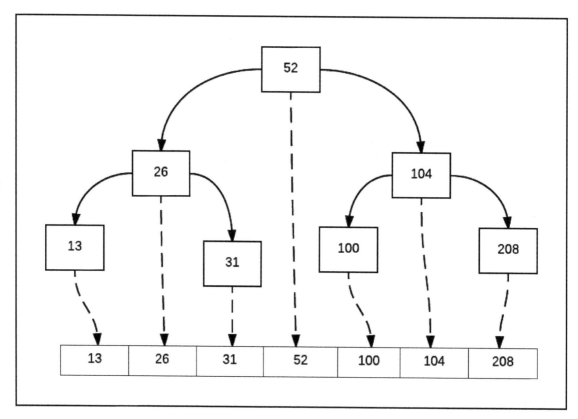

As you probably know, the visiting order is also the sorted order. If we put the values in an array as shown in the diagram, we get all the values sorted:

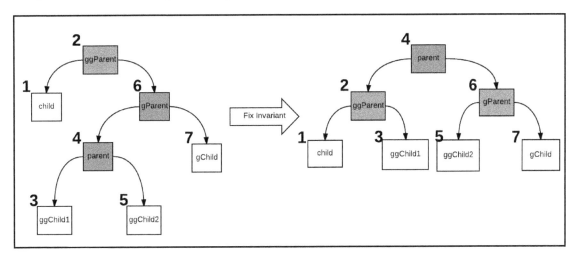

This traversal comes as a very useful tool to understand rotations. We can transform the tree, provided the **in-order traversal** visits the nodes in the same way.

The preceding diagram shows our right, left tree rotation. The numbers mark the order in which an in-order traversal would visit nodes.

The right-side-transformed tree also yields the nodes in the same order when traversed in an in-order fashion.

Drawing other rotations and verifying them in the same fashion will help you better understand the concept of rotations.

Complexity

What is the runtime complexity of node insertion? A Red-Black tree of n internal nodes has height at $2*log(n+1)$.

This means that operations, such as searching for a node, have logarithmic time. The insertion operation we just saw is also proportional to the height of the tree.

When we insert a new node and violate the second invariant, we fix it and *always* color the subtree's root *red*. This could create further invariant violation upward in the tree.

However, as the height is at most *2*log(n+1)*, the insertion operation has a runtime complexity of $O(log_n)$. So, for example, for a tree with *4294967296* nodes, which is 2^{32}, we have to perform up to 32 invariant fixes. This is a very good number, making Red-Black trees one of the most popular variants of Binary Search Trees.

Linux's completely fair scheduler uses Red-Black trees. See `http://www.ibm.com/developerworks/library/l-completely-fair-scheduler/` for more information.

Java 8's TreeMap is implemented using Red-Black trees. See `https://docs.oracle.com/javase/8/docs/api/java/util/TreeMap.html.`

Scala's TreeMap and TreeSet are also implemented using Red-Black trees.

Summary

We saw how a simple BST could degenerate into a linked list, for example, when we insert sorted data into the tree. In this case, instead of logarithmic lookup complexity, we should get a far slower, *O(n)* runtime complexity.

To make sure that the tree operations are logarithmic, we need to balance the tree. We learned about perfectly balanced trees, which are rare. Instead, height-balanced trees, which are almost balanced are good for us.

We touched upon some terms such as height of a node and internal nodes. Next, we looked at tree rotations, which are the basic building blocks for rebalancing a tree.

Red-Black trees are balanced BSTs, with every node colored in either red or black. There are two important invariants that need to be maintained.

We then had a detailed look at how inserting a node into a Red-Black tree keeps it balanced. Rebalancing is somewhat involved; however, we saw each case separately. Using in-order traversal, we showed how the rotations rebalance the tree without changing the traversal order.

Hope you enjoyed the ride! Let's continue the fun ride!

12
Binomial Heaps

In Chapter 8, *Queues* we looked at binary heaps. Now a binary *min-heap* takes the form of a complete binary tree. This means the key at each node is less than or equal to its children.

We will look at one more popular heap implementation, namely a binomial heap. A **binomial heap** is a collection of binomial trees, giving us a very efficient heap-merging operation.

We will begin with an introduction to binomial trees. Next, we will see how to link two binomial trees, the basics for growing a heap. The process of inserting into a binomial heap exhibits a surprising coincidence to the binary number addition process. This detour will help us understand the merge algorithm.

Next, we will look at how to merge two binomial heaps.

Finally, we will look at how to find and delete a minimum element. As we move on, we will reason the code and eventually exercise the various operations on REPL. We will do all this in good time, though. Let's review some basic terminologies before we look at the implementation:

- The depth of a node is the number of edges from the root to the said node
- The height of a node is the number of edges from the node to the bottommost leaf
- The height of the tree is the height of the root node

Here's an illustration of this:

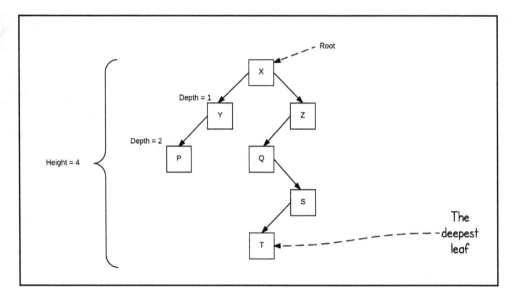

So here's a question for you, a quiz:

- What are the heights of the following trees?

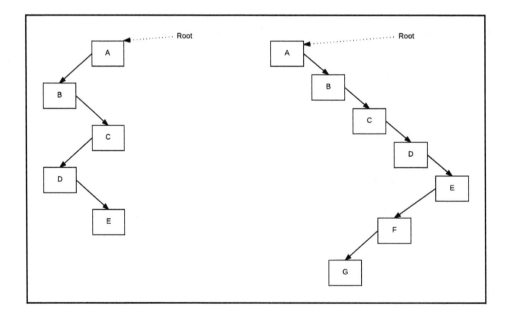

Binomial trees

What is a binomial tree? It is a recursively defined tree, as follows:

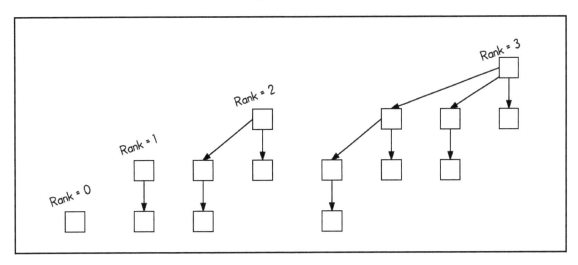

The first binomial tree has just one node. Its height is equal to *0*. A binomial tree of height *1* is formed from two binomial trees, each of height *0*. A binomial tree of height *2* is formed from two binomial trees, each of height *1*.

The tree is defined in terms of itself, recursively. For example, the following figure shows two binomial trees of rank *2*. When the right tree is pulled up, the left tree becomes its left child. The resulting tree is of rank 3 with $2^3 = 8$ nodes.

The structure is formed by similar smaller structures:

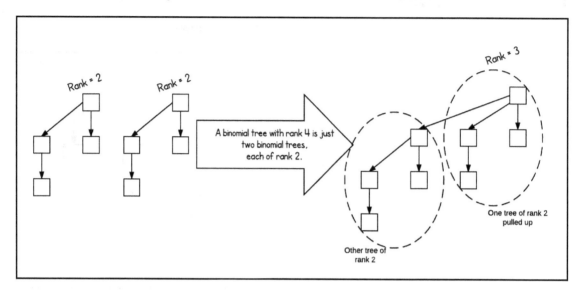

The definition of binomial trees could also be stated as follows: a *binomial tree with rank k is composed of subtrees with rank (k-1)*. Here is a pictorial way of stating this:

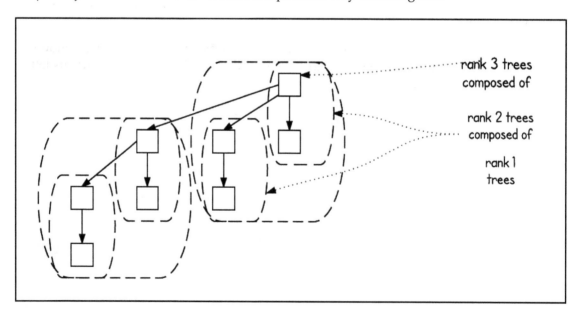

This forms an important basis for the *merging operation*, as we will soon see.

You can also look at a binomial tree as a list of binary trees. This is illustrated in the following figure:

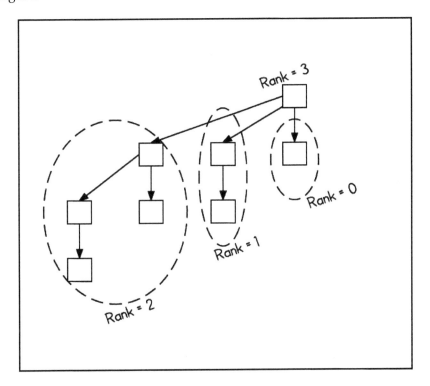

The figure shows a binomial tree of rank 3. If we leave out the top node, we have binomial trees of ranks 2, 1, and 0.

Left child, right sibling

How would we express the preceding data structure? For example, what is the canonical way of representing *n* children of a node? Consider you want to represent a tree. A node in a tree can have any number of children, and each child can have any number of children of its own:

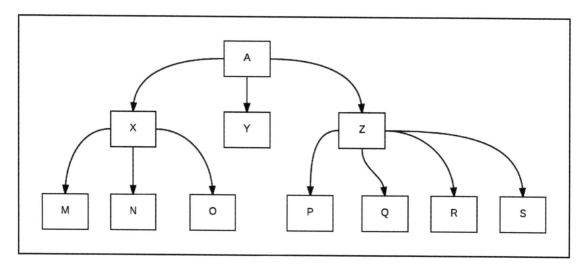

One way is for each node to hold a list of its children references. This is a straightforward representation. We wish to hold references to *n* children, and as there is no limit, we can also employ a list for holding the children. Here is how it would look pictorially:

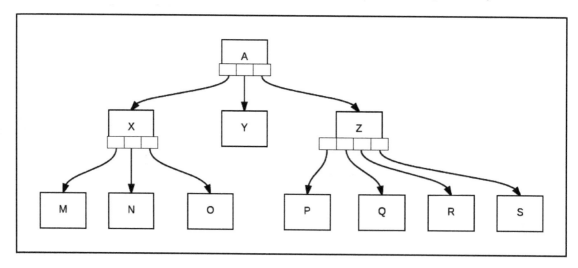

However, there is another representation, namely representing a tree as *a binary tree with the left pointer pointing to the first child and the right pointer to a sibling node*. Here is how it would look:

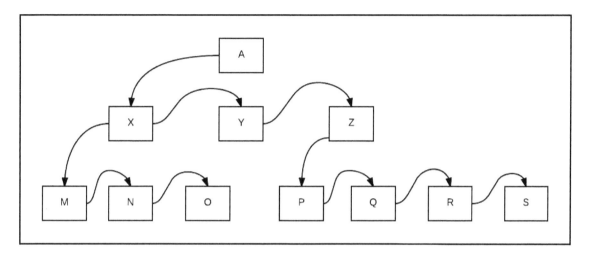

This is a better representation as we don't need to waste space for the array holding the children pointers.

We will use the first representation, though; can you guess why?

A binomial heap

A **heap-ordered binomial tree** is one in which every parent value is less than or equal to its children. In other words, a parent value *is never greater than its* children values.

Here's a diagram illustrating this:

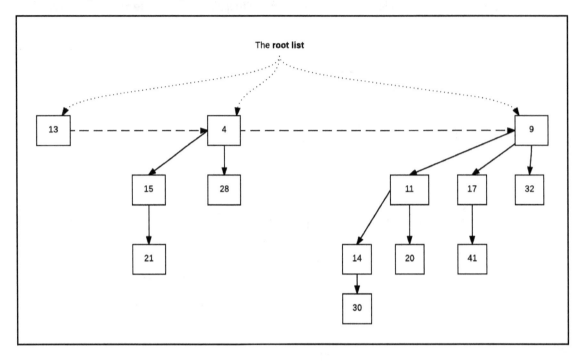

The diagram shows a binomial heap with **13** nodes. All the binomial trees are linked together in increasing order of their ranks. This linked list of roots is the *root list*.

Let's start shaping up our code now.

Linking up

Our node is defined as a `case` class:

```
case class Node(rank: Int, v: Int, children: List[Node])
```

The node holds the rank, the value v, and a list of children (possibly empty).

Given this definition, let's see how we could link two binomial trees. We always link trees of equal rank:

```
def linkUp(t1: Node, t2: Node) =
  if (t1.v <= t2.v)
    Node(t1.rank+1, t1.v, t2 :: t1.children)
  else
    Node(t1.rank+1, t2.v, t1 :: t2.children)
```

Here is the diagram showing linking up two trees of rank 0:

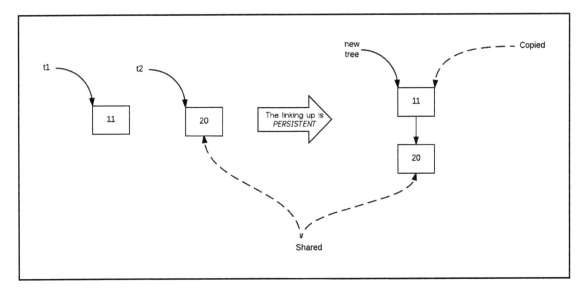

Let's grok this code with an example, for example, the case where both the trees are of rank 0, that is, each tree has only one node.

As t1 with value 11 is less than t2 (value 20), we copy t1 with an incremented rank. This copy of t1 becomes the new root.

The `linkUp` operation always holds the *heap property*. The parent value is *never greater than any child value*. The node `t2` is structurally shared. This structural sharing makes this a very fast operation. Here is another pictorial example where two trees of rank 2 are linked up:

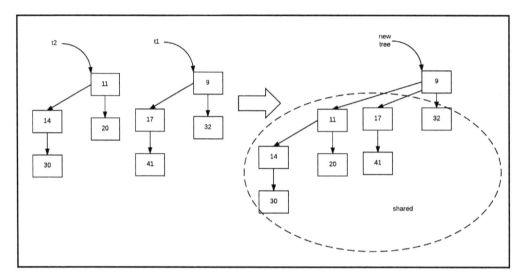

Most of the nodes of the original trees are shared, shown enclosed in a dashed circle.

A binomial heap *will never have two trees with an equal rank*. This is one invariant that needs to hold. The preceding `linkUp` routine is used to restore this invariant. We need to link up equally ranked trees in the `insert` and `merge` operations.

Inserting a value

Now that we have the linking up method under our belt, here is `insertElem`:

```
    def insert(t: Node, rootList: List[Node]): List[Node] = rootList match
{
        case Nil => List(t)
        case x :: xs if (t.rank < x.rank) => t :: rootList
        case x :: xs => insert(linkUp(t, x), xs)
    }

    def insertElem(rtList: List[Node], elem: Int) = {
        insert(Node(0, elem, Nil), rtList)
    }
```

The `insertElem` method takes the element to insert into a heap, which is a root list of binomial trees, as described in the preceding code. It uses a helper method, namely `insert`, to absorb the new element. Note that the `insert` method always returns a singleton tree, that is, a binomial tree of rank 0–a binomial tree with only one node:

```
case Nil => List(t)
```

When this clause matches, we get our very first node in the heap and the `rootList` is empty. We return a list with just one element, the argument:

```
case x :: xs if (t.rank < x.rank) => t :: rootList
```

In this case, we have found an *empty* slot:

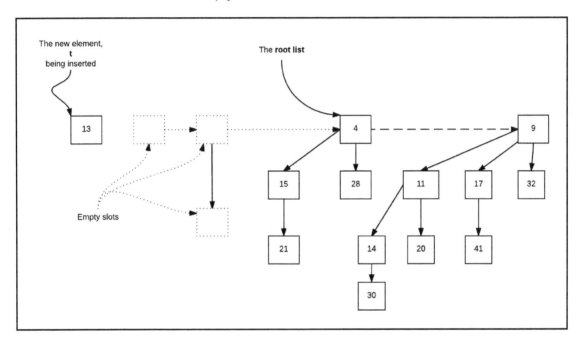

Here, we just prepend the new node. We know very well that prepending an element to a list is a very fast operation, that is, with complexity $O(1)$.

This process is very similar to adding 1 to the binary representation of some number, say, n. We will discuss this in more detail now.

The third clause is very interesting:

```
case x :: xs => insert(linkUp(t, x), xs)
```

Here we find that there is already a tree whose rank is equal to the tree being inserted. In this case, we link the two trees and insert the resulting tree back into the heap:

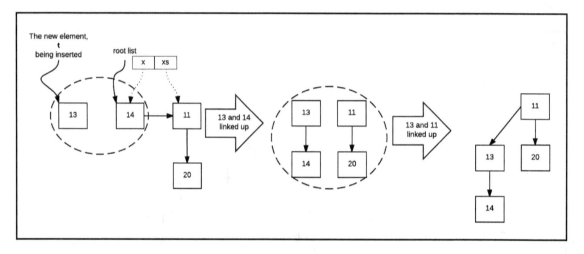

Note that the dotted area nodes are trees with equal ranks. The linking up can possibly result in more link ups and hence the recursive call to use the `insert` method again.

Binary number equivalence

There is a surprising equivalent in the processes of binomial heap insertion and incrementing a binary number. For example, the following figure shows the addition of 1 to a number, say 6, and the equivalent tree insertion into a binomial heap:

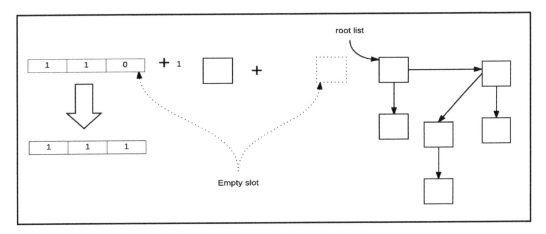

The binary addition happens from right to left, whereas binomial insertion happens from left to right. Also, the link up triggers changes to the heap similar to the way the carry triggers further changes to the number:

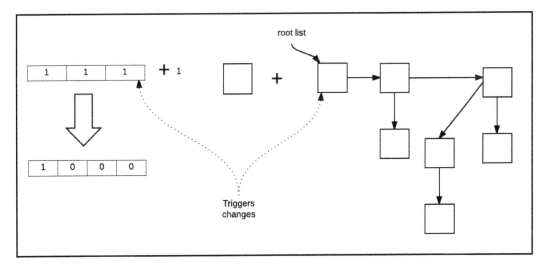

Merging

The `insert` method is a simplified case of the `merge` operation. We will discuss the case clauses one by one. This will help us see how the concepts detailed before fit together. Here are the first two cases:

```
    def merge(ts1: List[Node], ts2: List[Node]): List[Node] = (ts1, ts2)
match {
    case (ts1, Nil) => ts1
    case (Nil, ts2) => ts2
    case (t1 :: ts11, t2 :: ts22) if (t1.rank < t2.rank) => t1 ::
merge(ts11, ts2)
    case (t1 :: ts11, t2 :: ts22) if (t2.rank < t1.rank) => t2 ::
merge(ts1, ts22)
    case (t1 :: ts11, t2 :: ts22) => insert(linkUp(t1, t2), merge(ts11,
ts22))
    }
```

These two cases are pretty simple. Merging with a `Nil` list just returns the other list. Let's look at the third clause; the fourth is similar:

```
    case (t1 :: ts11, t2 :: ts22) if (t1.rank < t2.rank) => t1 :: merge(ts11,
ts2)
```

We look at the first nodes of both the lists. If the first node's rank is lesser, we prepend it to the result list. We also continue (recursively) with the *tail* of the first list and the entire second list.

The fourth case is similar. Working it out for yourself should be pretty easy.

The fifth case is simple too:

```
    case (t1 :: ts11, t2 :: ts22) => insert(linkUp(t1, t2), merge(ts11, ts22))
```

In the remaining cases, both the heads have equal ranks. We link them up and insert the resulting tree into the resulting list by merging both the tails.

Here's an exercise for you–work out the merge algorithm for the following trees:

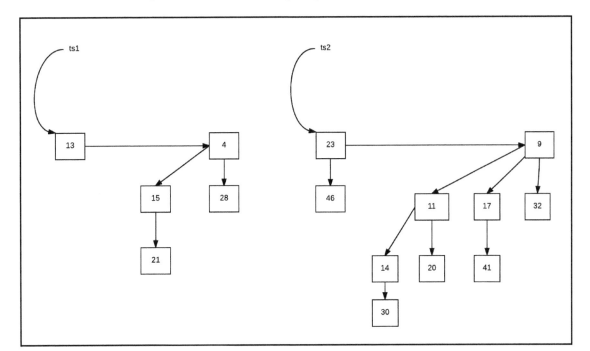

Find the minimum

A heap-ordered binomial tree holds the invariant that every tree node is smaller or equal compared to its children. Each element of the root list is a root of a heap-ordered binomial tree. So the minimum element of the *entire* heap is some element of the root list:

```
def findMin(rootList: List[Node]): Node = rootList min (Ordering.by((p:
Node) => p.v))
```

Dead simple, huh? We use the `min` method and pass in the `Ordering` criteria. Let's look at how the criteria works:

```
scala> case class Person(age: Int, name: String)
defined class Person

scala> val list = List(Person(10, "Sunil"), Person(13, "Dev"),
     |    Person(11, "Rajesh"))
list: List[Person] = List(Person(10,Sunil), Person(13,Dev),
Person(11,Rajesh))
```

```
scala> list.min(Ordering.by((p:Person) => p.age))
res0: Person = Person(10,Sunil)

scala> list.min(Ordering.by((p:Person) => p.name))
res1: Person = Person(13,Dev)
```

We have a `case` class with the properties `age` and `name`. The `Ordering` trait helps us define the ordering by either the name or age property. For more information on `Ordering`, see `http://www.scala-lang.org/api/current/index.html#scala.math.Ordering`.

Deleting the minimum

This is when we need to take the min element off the heap. There are three steps to do this:

1. Find the minimum element `min`.
2. Introduce a `newRootList` after removing `min`.
3. Merge `newRootList` and `min.children`.

The following diagram illustrates these steps:

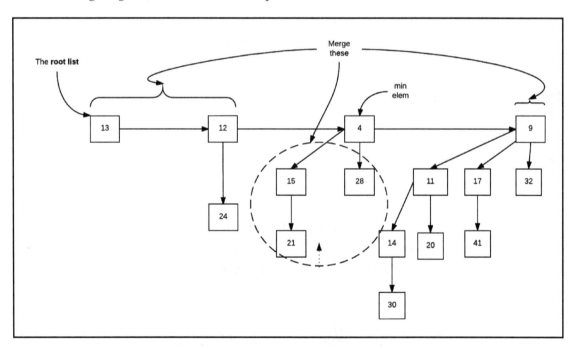

Here is the code:

```
def removeMin(rootList: List[Node]): List[Node] = {
  val min = findMin(rootList)
  val listExceptMin = rootList filterNot(_ == min)
  merge(listExceptMin, min.children)
}
```

This is a pretty neat algorithm. We find the minimum element and hold it in `min`. Next, we filter out `rootList` so everything, except the minimum element, is returned.

The `filterNot` method is the inverse of the `filter` method. Here is a quick Scala REPL session showing the difference:

```
scala> val list = List(1,2,3,4,5,6)
list: List[Int] = List(1, 2, 3, 4, 5, 6)
scala> list filter(_ % 2 == 0)
res3: List[Int] = List(2, 4, 6)
```

We find the even element by passing in a predicate function. To find the odd element, we use `filterNot`:

```
scala> list filterNot(_ % 2 == 0)
res4: List[Int] = List(1, 3, 5)
```

Once we remove the minimum, we need to take care of its children.

Exercising the code

Let's give it a go. The `insertElem` method is pretty handy:

```
scala> :load BinomialHeap.scala
Loading BinomialHeap.scala...
defined object BinomialHeap
scala> import BinomialHeap._
import BinomialHeap._
scala>    val v1 = insertElem(Nil, 11)
v1: List[BinomialHeap.Node] = List(Node(0,11,List()))

scala>    val v2 = insertElem(v1, 2)
v2: List[BinomialHeap.Node] = List(Node(1,2,List(Node(0,11,List()))))

scala>    val v3 = insertElem(v2, 9)
v3: List[BinomialHeap.Node] = List(Node(0,9,List()),
Node(1,2,List(Node(0,11,List()))))
```

```
scala> findMin(v3)
res1: BinomialHeap.Node = Node(1,2,List(Node(0,11,List())))
```

This is cumbersome, though. Note that `rootList` could be considered an accumulator of `Node` objects. We keep feeding it with numbers, each getting converted into a singleton binomial tree. You guessed it right! Scala's `foldLeft` method comes to the rescue:

```
scala>   val list = List(13, 12, 24, 4, 15, 28, 21, 9, 11, 17, 32, 14, 20,
41, 30)

scala>   val v = list.foldLeft[List[Node]](List[Node]())(insertElem)

scala>   println(v)
List(Node(0,30,List())...

scala>   println(findMin(v))
Node(3,4,List(...))
```

This gives us a very nice way to test the code for any combination.

While you are at it, it would help to try for a better string representation of the `Node` object. This is left as an exercise.

Complexity

There are four major operations: `insertElem`, `merge`, `findMin`, and `removeMin`. What is the complexity of these operations? We saw the equivalence of binomial heaps with a binary number. For example, inserting an element is so similar to adding 1 to a binary representation of some number, say, n.

How many elements does the root list have? It is proportional to $O(log_n)$. Because `insert` method keeps looking for an empty slot, for an almost full heap, the `insert` operation's complexity is $O(log_n)$.

Note that in many cases, we will hit an empty slot. Thus, the actual cost of an `insert` function is amortized: $O(1)$.

The `merge` operation has $O(log_n)$ time complexity.

Because we scan the root list for `findMin`, its complexity is $O(log_n)$. The `findMin` complexity could be optimized to $O(1)$. We could hold a reference to the minimum element and keep it updated when elements get inserted or removed.

The `deleteMin` operation's complexity is $O(log_n)$.

Summary

We looked at binomial heaps, which are heaps implemented using binomial trees. Binomial trees have a certain structure. They have an important property, a rank.

A binomial heap is composed of binomial trees, with the additional stipulation that every tree will have a distinct rank. The roots of a binomial tree are linked to a list in order of increasing rank, thereby forming a binomial heap. Every binomial tree in the heap adheres to the *heap property*.

We saw four major operations, namely `insertion`, `merge`, `findMin`, and `removeMin`. We saw how these operations have complexity of $O(log_n)$.

This is an interesting data structure and we invite you to compare it with the leftist heaps introduced earlier.

Let's continue the journey with a look at functional sorting algorithms.

13
Sorting

In a chaotic environment, we look for events that are ordered and follow some rules in order to understand them. An ordered environment also helps in searching and choosing elements according to our requirements for a specific purpose.

Sorting can be in an increasing order (not decreasing when the data list has duplicate elements), and it can be in a decreasing order too. Remember the queue of students in school in the increasing order of height in parades, sorting a deck of cards to get the required card in a shorter time while playing cards?

A sorting algorithm arranges the elements of a collection in some order, generally, in either increasing or decreasing order. A sorting algorithm requires the comparison of elements and putting them in a required order. So, a sorting algorithm involves comparison and swapping of elements. Whenever we think of the complexity of a sorting algorithm, we think in the context of the number of comparisons and number of swaps. If the number of comparisons or the number of swaps, or both, is decreased, algorithm's complexity also goes down.

A sorted collection is useful in some other algorithms, such as binary search, which searches data in a sorted collection or sequence.

Just consider a sequence that has n elements, $e1, e2, e3,..., en$. After sorting in an increasing order, the sequence of elements will have following order:

$e1 <= e2 <= e3 <= $ .... $ <= en-1 <= en$

If we sort in a decreasing order, then, after sorting, the sequence of elements will have the following structure:

$en => en-1 => en-2 => $ .... $ => e2 => e1$

The following is a pictorial representation of sorting in an increasing order:

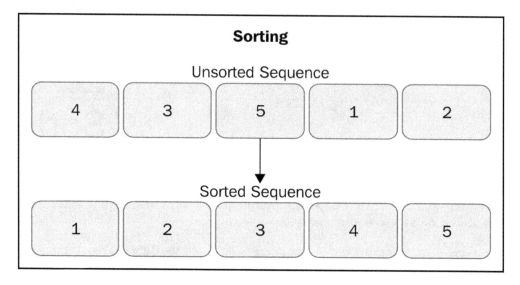

The efficiency of a sorting algorithm is measured as the time taken by that algorithm to sort N elements. One sorting algorithm might be more efficient than the other for same dataset; in other words, one algorithm is faster than the other.

In this chapter, we are going to discuss the following sorting algorithms:

- Bubble sort
- Selection sort
- Insertion sort
- Merge sort
- Quick sort

Stable and unstable sorting

In the following paragraphs, we will discuss stable and unstable sorting.

Stable sorting

A stable sort algorithm maintains the relative ordering of elements of equal values in a sorted sequence. It can be understood using the following diagram:

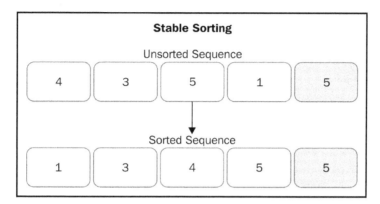

As the diagram depicts, our unsorted list has two fives. The first **5** is in a white slot and the second one is in a gray slot. After sorting, in the sorted sequence also, the **5** in the white slot remains before the **5** in the gray slot. This is an example of a stable sort.

Unstable sorting

Unstable sorting algorithms do not maintain the relative ordering of elements of equal values in a sorted sequence. The following diagram will help in understanding unstable sorting:

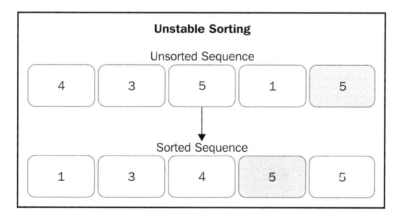

As shown in the figure, in a sorted sequence, the **5** in gray slot is before the **5** in white slot. In the unsorted sequence, the **5** in white slot is before the **5** in gray slot. After sorting, in the sorted sequence, their relative ordering is changed. This is an example of unstable sorting.

In the following sections, we will explore different sorting algorithms, their functional implementations, and their complexity analysis.

Bubble sort

Bubble sort is one of the simplest and oldest algorithms. Each element is compared with the next element. If neither of the elements are in order, then the elements are swapped. Every element of the sequence is visited.

A water bubble moves to the surface. Does this have any relation with bubble sort? The simplicity of the bubble sort algorithm makes it the starting point to learn a sorting algorithm. Bubble sort is stable as two elements of equal values are never swapped with each other.

We are going to implement all the algorithm in persistent way. You might be thinking that, what is the meaning of persistant data structure? Persistent data structure always maintains its previous version after modification. We are going to use another concept, that is structural sharing. In order to understand structural sharing consider following code example:

```scala
scala> val sl = List('c','e','g')
sl: List[Char] = List(c, e, g)
scala> val asl = 'a'::sl
asl: List[Char] = List(a, c, e, g)
scala> asl
res3: List[Char] = List(a, c, e, g)
```

We are creating a list `sl` having three elements. We have to prepend this list by a new element `a`. We are not changing the list `sl`, rather we are appending list `sl` to `a` and getting `asl`.

In a bubble sort, there are many passes. In each pass, we compare every element with its adjacent one and place them in order if they are not in order. If we are sorting in an increasing order, then in each pass, the largest element is moved to the last position. We leave that last element, find the biggest among the rest of the elements, and put it in the second last place.

In the last pass, all the elements get sorted. If we have to sort in a decreasing order, then we select the smallest elements by comparing every element and putting them in the last position.

The following figure shows the first pass of a bubble sort:

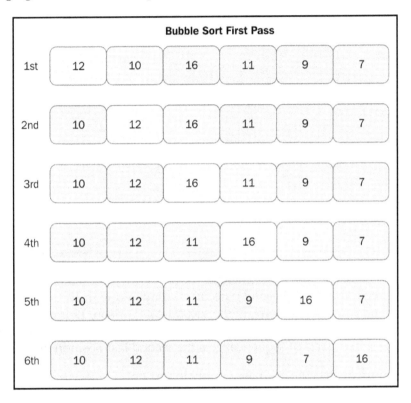

In the first step, elements **12** and **10** are compared. We want to sort in an increasing order. Since **12** is greater than **10**, they are swapped after comparison. In the second step, **12** and **16** are compared. Since **16** is greater than **12**, the elements are not swapped. The comparison and swapping is done in a similar way with the rest of the elements. In the last comparison, which is the fifth comparison, **7** is swapped with **16** and the largest element, which is **16**, is placed in the last position.

The following diagram depicts the second pass:

Bubble Sort Second Pass

1st	10	12	11	9	7	16
2nd	10	12	11	9	7	16
3rd	10	11	12	9	7	16
4th	10	11	9	12	7	16
5th	10	11	9	7	12	16

Before the second pass, the greatest element in the sequence has been placed in the last position. At the beginning of the second pass, **10** is compared with **12**. Since the data is already in order, no swapping is done. In the second comparison, **12** is compared with **11**. **11** is less than **12**. Therefore, the values are swapped. At the end of the second pass, **12** which is the largest in the rest of sequence, gets the second last position in the sequence.

The following figure depicts the last pass of bubble sorting:

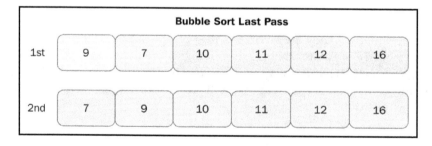

Bubble Sort Last Pass

1st	9	7	10	11	12	16
2nd	7	9	10	11	12	16

In last pass, **9** and **7** are compared and swapped. Finally, we get the sorted list of integers.

Scala implementation of bubble sort

In order to perform bubble sorting, we get the largest element in each pass at the end. The getLargest function gets the largest element of a sequence:

```scala
scala> def getLargest[T <% Ordered[T]](data : List[T]): (T,List[T]) = data
match {
  | case (Nil) => (null.asInstanceOf[T],Nil)
  | case (fv::Nil) => (fv,Nil)
  | case (fv::fl) => { val (fd, lso) = getLargest(fl)
  | if(fd >= fv)
  | (fd ,fv::lso)
  | else
  | (fv, fd :: lso)
  | }
  |
  | }
getLargest: [T](data: List[T])(implicit evidence$1: T => Ordered[T])(T,
List[T])
```

The getLargest function fetches the largest value from a sequence recursively. With the largest value in the sequence, it will also return the rest of the sequence. In the following part, we can see how the getLargest function fetches the largest value from a sequence. It is better to get some more insight about getLargest. Variable, fv, will get the first value of the list and variable fl will get the rest of the list. The fl variable is a list. If fl is null, then return it at the same moment. If fl is not empty, get the largest value out of fl recursively:

```scala
scala> val intData:List[Integer] = List(1,3,5,2,6)

intData: List[Integer] = List(1, 3, 5, 2, 6)

We have created a List of integers. 6 is largest value in the list.

scala> val largestInteger = getLargest(intData)

largestInteger: (Integer, List[Integer]) = (6,List(1, 3, 5, 2))

scala>  println(largestInteger)

(6,List(1, 3, 5, 2))
```

The getLargest function fetches the largest number from the intData list and also, the rest of the sequence. The rest of the sequence is in the same order as the mother list.

The following code snippet fetches the largest character from the list and the rest of the sequence:

```
scala>  val charData: List[String] = List("a","c","b","d")

charData: List[String] = List(a, c, b, d)

scala>      val largestCharacter = getLargest(charData)

largestCharacter: (String, List[String]) = (d,List(a, c, b))

scala>      println(largestCharacter)

(d,List(a, c, b))
```

The `getLargest` function can be modified to fetch the smallest number from the `List`.

Bubble sort, in the case of sorting in an increasing order, finds out the largest value and puts it in the last position of the unsorted part of the list in each pass.

The following code shows the final bubble sort implementation using the `getLargest` function:

```
scala> def bubbleSort[T <% Ordered[T]](data : List[T]) : List[T] = data
match {
     |        case Nil => Nil
     |        case _ =>
     |                 val (fv, sl) = getLargest(data)
     |                 bubbleSort(sl):::List(fv)
     |    }

bubbleSort: [T](data: List[T])(implicit evidence$1: T => Ordered[T])List[T]
```

The `bubbleSort` function fetches the largest value and the rest of the sequence using `getLargest`. Then, it recursively provides the full sorted list. In the following code snippet, we will finally see how the `bubbleSort` function provides a sorted list:

```
scala> val sortedIntegers  = bubbleSort(intData)

sortedIntegers: List[Integer] = List(1, 2, 3, 5, 6)

scala>      println(sortedIntegers)

List(1, 2, 3, 5, 6)
```

Using our implementation of bubble sort, we have got a sorted list of integers.

Our generalized bubble sort implementation also sort sequence of strings in increasing order as follows:

```scala
scala>  val sortedString  = bubbleSort(charData)

sortedString: List[String] = List(a, b, c, d)

scala>     println(sortedString)

List(a, b, c, d)
```

Complexity of bubble sort

Just consider that a sorting sequence has N elements. In the first pass, the algorithm has to compare $N-1$ elements. In the second pass, it performs $N-2$ comparisons. Similarly, in the third pass, the bubble sort algorithm performs $N-3$ comparisons. The last pass will have one comparison.

The overall complexity can be calculated as follows:

complexity = (N-1)+ (N-2) + (N-3) + ... + 3 + 2 + 1

$$= N(N-1)/2$$

Hence, the complexity of bubble sort is $O(N^2)$.

Selection sort

Selection sort is another simple sorting algorithm. It is a stable sorting algorithm. In bubble sort, we find that in every pass, if adjacent elements are not in order, they are swapped. Selection sort provides an improvement over bubble sort, with one swapping in every pass. In every pass, it finds out the largest or the smallest element and puts it in the right position. Consider that we have a sequence of integers.

Our sequence has [12,10,16,11,9,7] as elements. We can understand how selection sort works on this given sequence, with the help of the following diagram:

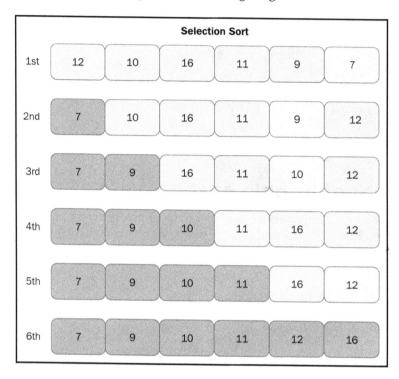

In the first pass of selection sort, the algorithm will look for the smallest element in the entire sequence. After finding the smallest element's location, it will swap it with the first element. In our given sequence, the smallest element in the first pass is **7**, which is in the last position in the sequence. After selecting this element, it will be swapped with the first element, **12**.

In the second pass, the smallest element will be selected from the rest of the sequence. The rest of the sequence possess [10,16,11,9,12]. The smallest element is **9** now and it will be swapped with **10**.

Similarly, in the fourth pass, half of the sequence is sorted, containing the elements **7**, **9**, and **10**. In the other part of the sequence, the smallest element, which is **11**, is selected. Element **11** is in the correct position; hence, it is not swapped with any element. In the same fashion, after the fifth pass, we get the list sorted in an increasing order.

So finally, we have understood selection sorting. Now, it is time to implement selection sort in Scala. The following code snippet shows the functional implementation of selection sort in Scala:

```
scala>  def selectionSort[T <% Ordered[T]](data : List[T]): List[T] = data
match {
        |       case Nil => Nil
        |       case fv::Nil => List(fv)
        |       case fv::fl =>
        |                       val minData = fl.min
        |                       val minIndex =  fl.indexOf(minData)
        |                       if(fv <= minData)
        |                         fv::selectionSort(fl)
        |                       else{
        |                        val (fp,sp)= fl.splitAt(minIndex)
        |                        sp.head::selectionSort(fp:::fv::sp.tail)
        |                       }
        |
        |    }
```

```
selectionSort: [T](data: List[T])(implicit evidence$1: T =>
Ordered[T])List[T]
```

The `selectionSort` function will perform the selection sorting for us. It is a recursive function. We will visualize this code by bisecting it:

```
case Nil => Nil
case fv::Nil => List(fv)
```

The preceding code snippet will return `Nil` if the sequence doesn't have any element. If the list has only one element, then the code snippet will return a list of one element.

The following code snippet will divide the list into two parts: `head` and `tail`:

```
case fv::fl =>
        |                       val minData = fl.min
        |                       val minIndex =  fl.indexOf(minData)
        |                       if(fv <= minData)
        |                         fv::selectionSort(fl)
        |                       else{
        |                        val (fp,sp)= fl.splitAt(minIndex)
        |                        sp.head::selectionSort(fp:::fv::sp.tail)
        |                       }
```

Then, in the `tail` part, it looks for the smallest element. Thereafter, the index of the smallest element is found. If the `head` element is less than the smallest element, then the `head` will not be swapped. The `tail` is sorted recursively.

However, if the `head` is not the smallest, then the list is split at the index containing the smallest number. The smallest number is swapped with the first element and the rest of list is again sorted recursively.

The following code snippet shows how to use the `selectionSort` function:

```
scala> val intData:List[Integer] = List(1,3,5,2,6)

intData: List[Integer] = List(1, 3, 5, 2, 6)
```

We have created a list of integers. Using the `selectionSort` function, we sort our list of integers, `intData`:

```
scala>    val sortedIntegers  = selectionSort(intData)

sortedIntegers: List[Integer] = List(1, 2, 3, 5, 6)

scala>    println(sortedIntegers)
List(1, 2, 3, 5, 6)
Now we will apply our selectionSort function on a list of Strings. For this
purpose, we have created a list charData of strings:scala> val charData:
List[String] = List("a","c","b","d")

charData: List[String] = List(a, c, b, d)


scala>    val sortedString  = selectionSort(charData)

sortedString: List[String] = List(a, b, c, d)

scala>    println(sortedString)

List(a, b, c, d)
```

Finally, we find that our functional implementation of selection sorting is working well with strings as well.

Complexity of selection sort

Just consider sorting a sequence of *N* elements. In the first pass, the algorithm has to compare *N-1* elements. In the second pass, it performs *N-2* comparisons. Similarly, in the third pass, selection sort performs *N-3* comparisons. The last pass will have one comparison.

The overall complexity can be calculated as follows:

complexity = (N-1)+ (N-2) + (N-3) + ... + 3 + 2 + 1

$$= N(N-1)/2$$

Hence, the complexity of selection sort is $O(N^2)$.

You might be thinking what will happen if we provide a sorted list. If we put a check to ascertain whether the list is already sorted, it requires $O(N)$ complexity. If the list is sorted, then just return it.

Insertion sort

Insertion sort is simple to implement and a stable sorting algorithm. During the process of sorting, it creates a sorted subsequence. From the unsorted subpart of the sequence, it takes one value at a time in each pass and inserts it in the sorted part of the sequence, maintaining the order in the sorted subsequence. When the last element is inserted in the sorted subsequence, we get the final sorted sequence. Hence, for a sequence of *N* elements, it takes *N-1* passes. Remember that it does not swap values like the bubble and selection sort algorithms, but it inserts values one by one in a sorted subsequence.

We can better understand the insertion sort mechanism with the following pass-by-pass diagram:

Insertion Sort

1st	12	10	16	11	9	7
2nd	10	12	16	11	9	7
3rd	10	12	16	11	9	7
4th	10	11	12	16	9	7
5th	9	10	11	12	16	7
last	7	9	10	11	12	16

We have a sequence of integers with elements, [12, 10, 16, 11, 9, 7]. In the first pass, it starts with **10**. The already sorted subsequence has one element, **12**. Element **10** is inserted in front of **12**. Now we have two elements, **10** and **12**, in the sorted subsequence.

We can visualize how things are done in the third pass from the following diagram:

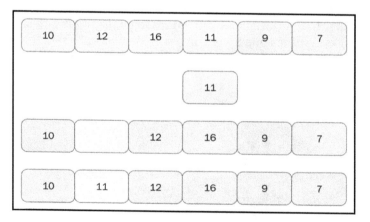

The element in the fourth position has to be inserted into the sorted subsequence. At that moment, our sorted subsequence has [10, 12, 16]. Element **11** is in the fourth position. Element **11** is compared with **10** and found greater than **10**. Now, **11** is compared with **12**; **11** is less than **12**. Hence, **11** is inserted between **10** and **12**. Now, the sorted subsequence has [10, 11, 12, 16].

After getting a clear understanding of the working of insertion sort, anyone will try to implement it in some programming language. In the following part of this chapter, we will explore the implementation of insertion sort in Scala in a functional way.

As we observed, in insertion sorting, an element is inserted in a sorted subsequence. So, first, we will observe the Scala method that will insert an element in the sorted subsequence:

```
scala>  def insertElement[T <% Ordered[T]](elm : T , sorted : List[T]):
List[T]= {
       |        if(sorted == Nil){
       |          return elm::sorted
       |        }
       |        val firstElement   :: tailElement = sorted
       |        if (firstElement < elm)
       |          firstElement :: insertElement(elm,tailElement)
       |        else
       |          elm :: sorted
       |    }

insertElement: [T](elm: T, sorted: List[T])(implicit evidence$1: T =>
Ordered[T])List[T]
```

The `insertElement` function will insert an element, `elm`, in the `sorted` list. It inserts `elm` in the `sorted` list recursively. In the `insertElement` function, the `sorted` list is broken into `head` and `tail`. The element that has to be sorted is compared with the `head` element. If `elm` is greater than the `head` element, it is inserted in rest of sorted subsequence recursively.

The following code snippet will make us very clear about the output of the `insertElement` function for the given input:

```
scala>  val ls : List[Integer] = List(1,4,5)

ls: List[Integer] = List(1, 4, 5)

scala>      val v : Integer= 3

v: Integer = 3

scala>      val insertedList =  insertElement(v,ls)
```

```
insertedList: List[Integer] = List(1, 3, 4, 5)

scala>        println(insertedList)

List(1, 3, 4, 5)
```

We have created a list, `ls`, of integers with elements 1, 4, 5. We can observe that `ls` is sorted in an increasing order. The integer element 3 has to be inserted in the list, `ls`. The `insertElement` function is used and we finally get `insertedList`. The `insertedList` list has element 3 inserted in the appropriate position.

So, we have understood the Scala function, `insertElement`, and how it inserts a given element in a sorted list. Now is the time when the full insertion sort algorithm should be implemented using the `insertElement` function:

```
scala> def insertionSort[T <% Ordered[T]](data : List[T]) : List[T] = {
     |        if(data == Nil)
     |            data
     |        else{
     |            val firstElement = data.head
     |            val tailElement = data.tail
     |            val temp  = insertionSort(tailElement)
     |            insertElement(firstElement, temp)
     |        }
     |    }

insertionSort: [T](data: List[T])(implicit evidence$1: T =>
Ordered[T])List[T]
```

The `insertionSort` function will take a sequence and give a sorted sequence as the output. It is a functional implementation. Here, insertion sort is done recursively:

```
scala>   val intData:List[Integer] = List(1,3,5,2,6)

intData: List[Integer] = List(1, 3, 5, 2, 6)

scala>        val sortedIntegers  = insertionSort(intData)

sortedIntegers: List[Integer] = List(1, 2, 3, 5, 6)

scala>        println(sortedIntegers)

List(1, 2, 3, 5, 6)

scala>   val charData: List[String] = List("a","c","b","d")

charData: List[String] = List(a, c, b, d)
```

```
scala>        val sortedString  = insertionSort(charData)

sortedString: List[String] = List(a, b, c, d)

scala>        println(sortedString)

List(a, b, c, d)
```

In the preceding code snippet, we saw that our insertion sort implementation in Scala works fine both on integers and characters.

Insertion sort can perform online sorting. So, what is the meaning of online sorting? Just consider that we have a small list, [3, 1, 2]. We sort it and get the sorted list, [1, 2, 3]. After some time, we get some more data [5, 3, 4]. We have to merge these two lists in a sorted fashion. Using insertion sort, we can do it easily.

Complexity of insertion sort

Consider that a sorting sequence has *N* elements. In the first pass, the algorithm has to compare one element. In the second pass it has to perform two comparisons. Similarly, in the third pass, selection sort performs three comparisons. The last pass will have *N-1* comparisons.

The overall complexity can be calculated as follows:

complexity = 1 + 2 + 3 + ... + (N-3) + (N-2) + (N-1)

$$= N(N-1)/2$$

Hence, the worst complexity of selection sort is $O(N^2)$.

You might be thinking what will happen if we provide a sorted list? If we put a check to ascertain whether the list is already sorted, it requires *O(N)* complexity. If the list is sorted, then it will just return it.

Merge sort

Merge sort uses the *divide and conquer* philosophy. There will be a function split that will split a sequential data structure into two parts in such a way that the number of elements in the two split parts will differ by one element at the maximum. The split daughter lists, after merging, return the permutation of the original list.

Merge sort steps can be laid out in the following fashion:

1. The sequence is split till every subsequence has at most one element.
2. Every subsequence is merged in such a way that the merged sequence is sorted too.

No worries, we will discuss all the steps in detail in the following pages. Let's start with the split.

Splitting the sequence

The following diagram explains the splitting of a sequence:

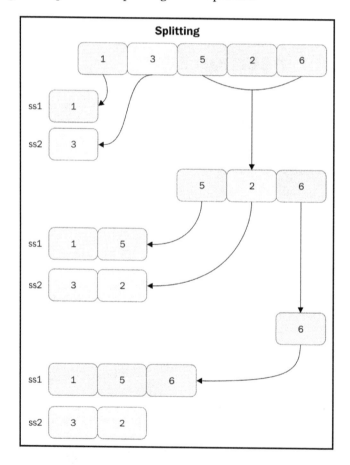

We start our sequence as [1, 3, 5, 2, 6]. In order to split it into two parts, we will fetch two elements (they are **1** and **3** for us) and put them as two subsequences, **ss1** and **ss2**, respectively. Similarly, the next time, **5** and **2** will be taken and added to subsequences **ss1** and **ss2**, respectively. At last, **6** is added to the first subsequence, that is, **ss1**. All these steps are performed recursively to maintain our functional programming style.

In following code snippets, we will discuss the partition in Scala:

```scala
scala>  def split[T <% Ordered[T]](fl : List[T]) :(List[T], List[T]) = {
     |      if(fl == Nil){
     |         return (Nil,Nil)
     |      }
     |    val fv = fl.head
     |    val ls1 = fl.tail
     |    if (ls1 == Nil) {
     |         return(fv::Nil , Nil)
     |    }
     |    val sv = ls1.head
     |    val ls2 = ls1.tail
     |    val(tl1, tl2) = split(ls1.tail)
     |    return(fv::tl1, sv::tl2)
     |  }

split: [T](fl: List[T])(implicit evidence$1: T => Ordered[T])(List[T],
List[T])
```

We saw how the splitting is done for merge sorting. We can observe how the first and the second element of a sequence are taken out. The fetched elements will be added to two different subsequences, respectively:

```scala
scala>  val intData:List[Integer] = List(1,3,5,2,6)
intData: List[Integer] = List(1, 3, 5, 2, 6)

scala>      val (firstList, secondList) = split(intData)
firstList: List[Integer] = List(1, 5, 6)
secondList: List[Integer] = List(3, 2)

scala>      println(firstList)
List(1, 5, 6)

scala>      println(secondList)
List(3, 2)
```

We can observe how out function split divides our integer list `intData` into two parts. Since it is implemented in a generic way, it also works on strings:

```
scala> val lst = List(5,3,4,2,1)
lst: List[Int] = List(5, 3, 4, 2, 1)

scala> val splitLength = lst.length/2
splitLength: Int = 2

scala> val (fl, ll) =  lst.splitAt(splitLength)
fl: List[Int] = List(5, 3)
ll: List[Int] = List(4, 2, 1)
```

Splitting the list can also be done using the `splitAt` function.

Merging two sorted subsequences

Two sorted subsequences can be merged to get a sorted sequence. The following pictorial representation throws proper light on the solution:

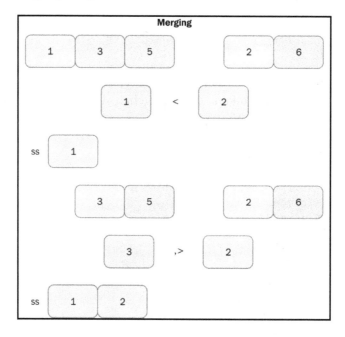

As the preceding image depicts, we have two sorted subsequences. The first one has [1, 3, 5], and the second subsequence's elements are [2, 6]. First, get the first element of each subsequence, which are **1** and **2**. 1 is less than 2; hence, in the new subsequence **1** will be added. Similarly, the next time, **3** and **2** will be the candidates for comparison. Element **2** is smaller and is added behind **1** in the output sequence. Remember that all these things are done here in a functional way. Following the same path, we get our subsequences merged in a sorted manner.

The understanding of an algorithm requires implementation to get more confidence. Let's see the implementation of merge and discuss. As a learner, you are always encouraged to come up with more efficient implementations than this book:

```scala
scala> def merge [T <% Ordered[T]] ( x: List[T], y : List[T]) : List[T] =
(x,y) match {
    |
    |        case (x, Nil) => x
    |        case (Nil,y) =>  y
    |        case( fm::fl, sm::sl) => if (fm > sm) sm::merge(x,sl)
    |                                 else fm::merge(fl,y)
    |    }


merge: [T](x: List[T], y: List[T])(implicit evidence$1: T =>
Ordered[T])List[T]
```

The `merge` function will take two sorted lists and provide a merged list. The output list will be in a sorted fashion too. It is a recursive function. It fetches one element from each of the two subsequences and compares them. It then adds the smaller element to the output sequence and repeats the same recursively.

We have seen the splitting of a list and the merging of two sorted lists. How do we use these two to complete our merge sort? Let's check out the pictorial representation of merge sort:

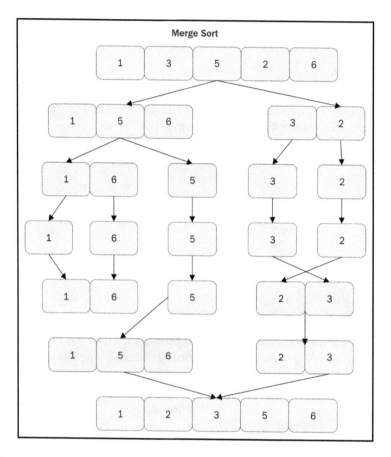

Let's start with an unsorted sequence having elements [1, 3, 5, 2, 6]. For merge sort, we have to first split it into smaller subsequences. Recursively, it is broken till every subsequence has one element. Then the subsequences are merged to get another sorted subsequence. **1** and **6** are merged to get another sequence. Remember, that merging is done in such a way that the elements of the resulting subsequence must be in a sorted fashion. All these intermediate subsequences are merged again to get the final sorted sequence.

Understanding the merge sort algorithm is done. How about a discussion on the implementation of merge sort in Scala? The following code is a functional implementation of merge sort in Scala:

```
scala> def mergeSort [T <% Ordered[T]] (ls :List[T]): List[T] = {
     |      if( ls == Nil || ls.tail == Nil)
     |        return ls
     |
     |      val (fl,sl) = split(ls)
     |      val lsm1 = mergeSort(fl)
     |      val lsm2 = mergeSort(sl)
     |      merge(lsm1, lsm2)
     |   }
mergeSort: [T](ls: List[T])(implicit evidence$1: T => Ordered[T])List[T]
```

The `mergeSort` function will do a merge sort on the sequence. The sequence is split and then the merge sort is applied on each subsequence:

```
scala>   val sortedIntegers   = mergeSort(intData)

sortedIntegers: List[Integer] = List(1, 2, 3, 5, 6)

scala>        println(sortedIntegers)

List(1, 2, 3, 5, 6)
```

In the preceding code snippet, `mergeSort` is being performed on a list of integers. After merge sort, we get the sorted list.

The `merge` function works fine with strings also:

```
scala> val firstSortedListChar: List[String] = List("a","c")

firstSortedListChar: List[String] = List(a, c)

scala>        val secondSortedListChar : List[String] = List("b","d")

secondSortedListChar: List[String] = List(b, d)

scala> val mergedListChar :List[String] =
merge(firstSortedListChar,secondSortedListChar)
mergedListChar: List[String] = List(a, b, c, d)

scala>println(mergedListChar)

List(a, b, c, d)
```

Complexity of merge sort

The average and worst case complexity of merge sort is $O(nlogn)$. Let me explain how I reached this conclusion. Look at every step. There are two steps. The first step is breaking a list, which takes $O(n)$ complexity, and then, merging the list, which again takes $O(n)$ complexity. So, the total is $O(2n)$ complexity, which is equivalent to $O(n)$. All these steps go to a height of $log\ n$. Therefore, the total complexity of merge sort is $O(nlogn)$. For a more detailed study of merge sort complexity calculation go through
http://math.stackexchange.com/questions/54416/merge-sort-time-complexity-analys is.

The following link is also very useful:

http://math.stackexchange.com/questions/707894/recurrence-relation-and-big-o-no tation

Quick sort

Quick sort is also known as *partition exchange sort*. It was developed by Tony Hoare. Quick sort also utilizes the power of *divide and conquer*, which we came to understand in the previous section. It is just about dividing a problem into approximately similar sub problems, solving each sub problem separately, and combining the results of the sub problems to deliver the final result.

Quick sort steps can be laid out in following fashion:

1. We select a pivot element. Selecting a proper pivot is very necessary for efficiency in a quick sort algorithm.
2. We put the pivot element in such a location that all the elements on its left are less than the pivot element, and all the elements on its right are greater than the pivot. This way, we partition the sequence into two parts.
3. Sort the subsequence on the left side of the pivot element and that on the right side of the pivot element recursively.

No worries, we will have a detailed discussion on all the steps in the following pages. Let's start with the partition.

Partition

Consider an integer sequence [4, 3, 5, 2, 6]. We have to partition it into two subsequences using a pivot element. There are many ways of getting a pivot element, such as generating a random number between **1** and **5**. Let the randomly generated number be **3**, then select **5** as the pivot element, which is in the third location. But we will go simply. Let **4**, which is the first element of the sequence, be the pivot element. Then, out of the other elements, [3, 5, 2, 6], **3** and **2** are less and **5** and **6** are greater than **4**. So, using this pivot element, we will get the two subsequences. The first subsequence has [2, 3] and the second subsequence has [5, 6]. The partition can be understood in a clearer way with the following pictorial representation of the partition:

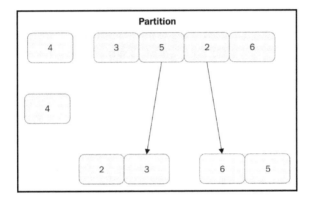

The following is the implementation of the partition in Scala:

```
scala>  def partition[T <% Ordered[T]](elm : T, seq : List[T], fp :List[T],
sp: List[T]):
                                                              (List[T],
List[T]) = seq match {
    |        case fe::fl => if (fe < elm) partition(elm, fl, fe::fp,sp)
    |                                 else
    |                                     partition(elm,fl,fp,fe::sp)
    |        case Nil => (fp,sp)
    |
    |    }
  partition: [T](elm: T, seq: List[T], fp: List[T], sp: List[T])(implicit
evidence$1: T => Ordered[T])(List[T], List[T])
```

The partition function will take the pivot element, elm, and a list to be partitioned. It will provide two partitioned lists as the output. The first partition will have all the elements that are less than the pivot element. The second partition will have all the elements greater than the pivot element.

The following code snippet shows how to use the `partition` function:

```scala
scala>   val intData1:List[Integer] = List(3,5,2,6)

intData1: List[Integer] = List(3, 5, 2, 6)

scala> val elm : Integer= 4
elm: Integer = 4

scala>   val (firstList, secondList) = partition(elm,intData1,Nil, Nil)

firstList: List[Integer] = List(2, 3)
secondList: List[Integer] = List(6, 5)
```

The preceding code snippet partitions the `intData1` list with the pivot element, 4. After the partitioning, it provides `firstList` and `secondList`. The `firstList` has 2 and 3, which are less than 4. `secondList` has elements 6 and 5, which are greater than 4.

Let's move our discussion forward to quick sort and its implementation. As we discussed earlier, quick sort is a *divide and conquer* algorithm. A pictorial representation of quick sort is as follows:

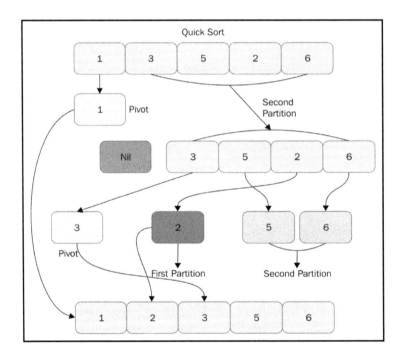

In order to explain quick sort, let's start with a sequence [1, 3, 5, 2, 6]. The first pivot element is 1, which is already the smallest number in the sequence. So, after the first partitioning, the first partition is **Nil** and the second partition has [3, 5, 2, 6]. For the second pass, **3** becomes the pivot element. After splitting, the first partition has **2** and the second partition has [5, 6]. In every split, one element gets to its sorted location. Finally, we get a sorted list.

For more information on complexity analysis of algorithms please refer to, `http://stackov` `erflow.com/questions/2695066/analysis-of-algorithms-complexity`

Our quick sort discussion now moves on to its implementation in Scala. The following is an implementation of quick sort in Scala:

```scala
scala> def quickSort[T <% Ordered[T]](ls : List[T]): List[T] = ls match {
     |      case Nil => Nil
     |      case fv:: Nil => List(fv)
     |      case fv::fl => {
     |                          val (l1, l2) = partition( fv,fl, Nil,Nil)
     |                          val lfl = quickSort(l1)
     |                          val rhl = quickSort(l2)
     |                          val tmp = fv :: rhl
     |                          return lfl ++ tmp
     |                     }
     |   }

quickSort: [T](ls: List[T])(implicit evidence$1: T => Ordered[T])List[T]
```

As we can see, after getting a sequence, the first element is taken as the pivot. Then, the rest of the sequence is partitioned, and then, further partitioned subsequences are sorted recursively:

```scala
scala> val intData:List[Integer] = List(1,3,5,2,6)
intData: List[Integer] = List(1, 3, 5, 2, 6)

scala>       val sortedIntegers  = quickSort(intData)
sortedIntegers: List[Integer] = List(1, 2, 3, 5, 6)

scala>       println(sortedIntegers)
List(1, 2, 3, 5, 6)
```

Fabulously, our implementation function, `quicksort`, has sorted integers in the best way. Will it work with strings? Let's see:

```scala
scala> val charData: List[String] = List("a","c","b","d")
charData: List[String] = List(a, c, b, d)

scala>       val sortedString  = quickSort(charData)
sortedString: List[String] = List(a, b, c, d)
```

```
scala>       println(sortedString)
List(a,  b,  c,  d)
```

As we guessed, the `quickSort` function works appropriately with strings too because it is a generic implementation.

Complexity of quick sort

The worst case complexity of quick sort is $O(n^2)$. The average case complexity of quick sort is $O(nlogn)$. Let's start discussing the average case. Let, every time, the pivot element be placed in the middle of the list. In this condition, as we mentioned before, the depth it will move is $logn$, and in every step, the complexity is $O(n)$. Hence, the total complexity is $O(nlogn)$.

Let's discuss the worst case scenario. What are you thinking? What will be the worst case scenario? Just consider a scenario where you are providing an already sorted list. Now, the tree's height will be n. Every step's complexity is $O(n)$. Hence, the total complexity is $O(n^2)$.

Summary

Sorting is used to solve many complex problems. Sorted data aids searching algorithms too. We discussed some very useful sorting algorithms, such as bubble sort, selection sort, insertion sort, merge sort, and quick sort. Every sorting algorithm has its own complexity.

We discussed bubble sort. Bubble sort is one of the earliest developed sorting algorithms. We implemented it in Scala.

After bubble sort, we moved on to selection sort, where in every pass, a small element is selected and put in a sorted list. Here, we found that in every pass, only one exchange is done, but the number of comparisons will be the number of unsorted elements.

The simplicity of insertion sort touched everyone. It uses two subsequences: one sorted and one unsorted. We take an element from the unsorted subsequence and put it in the sorted subsequence, while maintaining the order of the elements.

Divide and conquer is the most celebrated technique for problem solving. It has been used in sorting algorithms. Merge sort and quick sort are good example of *divide and conquer* algorithms. We discussed merge sort and its implementation in Scala. We also saw the quick sort algorithm and its implementation.

Index